ILSI INTERNATIONAL LIFE SCIENCES INSTITUTE

ILSI Human Nutrition Reviews

Series Editor: Ian Macdonald

Already published:

Sweetness
Edited by John Dobbing

Calcium in Human Biology
Edited by B. E. C. Nordin

Sucrose: Nutritional and Safety Aspects
Edited by Gaston Vettorazzi and Ian Macdonald

Zinc in Human Biology
Edited by C. F. Mills

Dietary Starches and Sugars in Man: A Comparison
Edited by John Dobbing

Diet and Behaviour: Multidisciplinary Approaches
Edited by G. H. Anderson, N. A. Krasnegor, G. D. Miller and A. P. Simopoulos

Forthcoming in the series:

Thirst: Physiological and Psychological Aspects
Edited by D. A. Booth and D. J. Ramsay

Dietary Fibre
Edited by T. F. Schweizer, C. E. Edwards and M. A. Eastwood

Modern Lifestyles, Lower Energy Intake and Micronutrient Status

Edited by Klaus Pietrzik

With 39 Figures

Springer-Verlag
London Berlin Heidelberg New York
Paris Tokyo Hong Kong

Editor
Prof. Dr. Klaus Pietrzik
University of Bonn, Institute of Nutritional Science,
Dept. Pathophysiology of Human Nutrition,
Endenicher Allee 11-13,
D-5300 Bonn 1, West Germany

Series Editor
Ian Macdonald, MD, DSc, PhD, FIBiol
Emeritus Professor of Applied Physiology,
University of London, UK

ISBN 978-1-4471-3272-1 ISBN 978-1-4471-3270-7 (eBook)
DOI 10.1007/978-1-4471-3270-7

British Library Cataloguing in Publication Data
Pietrzik, Klaus
Modern lifestyles, lower energy intake and micronutrient status
1. Man. Health. Effects of diet
I. Title II. Series
613.2
ISBN 978-1-4471-3272-1

Library of Congress Cataloging-in-Publication Data
Modern lifestyles, lower energy intake, and micronutrient status/edited by Klaus Pietrzik
p. cm – (ILSI human nutrition reviews)
Includes index.
ISBN 978-1-4471-3272-1
1. Malnutrition. 2. Food habits. 3. Trace elements in nutrition.
I. Pietrzik, K. (Klaus) II. Series
RA645.N87M63 1991 90-42650
612.3–dc20 CIP

Softcover reprint of the hardcover 1st edition 1991

Typeset by Nuts & Muttons Typesetting Ltd, Capricorn House, Linton, Cambridgeshire

2128/3830–543210 Printed on acid-free paper

Foreword

This volume is one of a series concerned with topics considered to be of growing interest to those whose ultimate aim is the understanding of the nutrition of man. Volumes on Sweetness, Calcium in Human Biology, Zinc in Human Biology, Dietary Starches and Sugars in Man, Sucrose: Nutritional and Safety Aspects, Diet and Behaviour: Multidisciplinary Approaches have been published and another on Thirst is in preparation.

Written for workers in the nutritional and allied sciences rather than for the specialist, they aim to fill the gap between the textbook on one hand and the many publications addressed to the expert on the other. The target readership spans medicine, nutrition and the biological sciences generally and includes those in the food, chemical and allied industries who need to take account of advances in those fields relevant to their products. Funded by industry but with independent status, the International Life Sciences Institute (ILSI) is a non-profit organization founded to deal objectively with the numerous health and safety issues that today concern industry internationally. ILSI sponsors scientific research, organises conferences and publishes monographs relative to these problems.

This volume presents the results of the contributions and discussions of a workshop which was organized by the Working Group on Nutrition of ILSI Europe, a branch of the International Life Sciences Institute (ILSI).

London
April 1990

Ian Macdonald
Series Editor

Preface

Lifestyles continuously undergo changes. In the last decades, in particular, objective data have become available about these changes: they suggest that, among other factors, there is a reduction in energy intake, which is leading to critical micronutrient status in some population groups in Western Europe.

Severe nutritional deficiencies, such as scurvy and beri beri, are not found in Western Europe, but anaemia, goitre and other ailments are common. One result of changes in modern lifestyles is that one quarter of all illnesses from nutritional deficiencies are due, partly at least, to micronutrient malnutrition.

There has been very little consideration of how to deal with nutritionally related health problems in Western Europe, and two reasons may be advanced to explain this. In the first place, it is hard to believe that deficiencies arise in the midst of overconsumption and, secondly, there are difficulties in estimating the milder degrees of malnutrition.

A workshop was, therefore, held in Mallorca from 2 to 4 March 1989, to consider present knowledge of the relation of energy intake to micronutrient status. Specialists with international reputations in different disciplines – epidemiologists, nutritionists, chemists and medical practitioners – presented their data on how micronutrient intakes differed with lifestyle, and discussed the functional significance of marginal deficiencies of micronutrients. The meeting was closed, and only invited speakers took part but, because of the high quality of the papers and the conclusions and recommendations that resulted from the discussions, it has now been decided to publish the material presented to make it available to a wider readership.

I should like to thank ILSI Europe, who organized the workshop and Professor Ian Macdonald, the Series Editor of Human Nutrition Reviews, for his help and advice. I am also grateful to all the authors whose cooperation made this volume possible, and to the publishers for their assistance.

Bonn
March 1990

Klaus Pietrzik

Contents

Contributors

Dr. O. Amédée-Manesme
INSERM U 56 et Département de Pédiatrie, Hôpital de Bicetre, Kremlin-Bicetre 94275, France

Dr. W. Bayer
University of Tübingen, Äussere Weiler Strasse 12, 7413 Gomaringen, Germany

Prof. G. Brubacher
Institute of Biochemistry, University of Basle, Vesalgasse 1, 4051 Basle, Switzerland

Dr. R. Buzina
Institute of Public Health, Zagreb, and Medical School of the University of Zagreb, Yu-41000 Zagreb, Yugoslavia

Dr. C. Carlier
INSERM U 56 et Département de Pédiatrie, Hôpital de Bicetre, Kremlin-Bicetre 94275, France

Dr. J. Durlach
Société internationale pour le développement des recherches sur le magnesium, 2, rue de Villersexel, 75007 Paris, France

Prof. F. Fidanza
Istituto di Scienza dell'Alimentazione, Universita degli Studi di Perugia, Casella Postale 333, I-06100 Perugia, Italy

Prof. K. F. Gey
Vitamin Research Department, F. Hoffman-La Roche & Co. Ltd, CH-4002 Basle, Switzerland

Prof. L. Hallberg
University of Goteborg, Department of Medicine, 11, Sahlgrenska Sjukhuset, S-41345 Goteborg, Sweden

Dr. E. Hermann-Kunz
Institut für Sozialmedizin und Epidemiologie des Bundesgesundheitsamtes, Postfach 33 OD 13, D-1000 Berlin 33, Germany

Prof. A. Kallner
Department of Clinical Chemistry, Karolinska Hospital, S-1041 Stockholm, Sweden

Dr. L. Kohlmeier
Institut für Sozialmedizin und Epidemiologie des Bundesgesundheitsamtes, Postfach 33 OD 13, D-1000 Berlin 33, Germany

M. J. P. Mareschi
Groupe BSN, 7, Rue de Teheran, F-75381 Paris, Cedex 08, France

Dr. G. Mensink
Institut für Sozialmedizin und Epidemiologie des Bundesgesundheitsamtes, Postfach 33 OD 13, D-1000 Berlin 33, Germany

Dr. M. S. Mourey
INSERM U 56 et Département de Pédiatrie, Hôpital de Bicetre, Kremlin-Bicetre 94275, France

Prof. K. Pietrzik
Institute of Nutritional Science, Dept. Pathophysiology of Human Nutrition, University of Bonn, Endenicher Allee 11–13, D–5300 Bonn 1, Germany

Dr. A. Prentice
Medical Research Council Dunn Nutrition Unit, Downhams Lane, Milton Road, Cambridge CB4 1XJ, UK

Dr. B. Sandstrom
Research Department of Human Nutrition, Rolighedsvej 25, DK-1958 Frederiksberg C, Denmark

Prof. K. H. Schmidt
University of Tübingen, Äussere Weiler Strasse 12, 7413 Gomaringen, Germany

Dr. J. Schrijver
TNO-CIVO Toxicology and Nutrition Institute, Department of Clinical Biochemistry, PO Box 360, NL-3700 AJ Zeist, The Netherlands

Dr. K. Suboticanec
Institute of Public Health, Zagreb, and Medical School of the University of Zagreb, Yugoslavia

Dr. H. van den Berg
Department of Clinical Biochemistry, Institute TNO/CIVO, PO Box 360, NL-3700 AJ Zeist, The Netherlands

Prof. R. G. Whitehead
Medical Research Council Dunn Nutrition Unit, Downhams Lane, Milton Road, Cambridge CB4 1XJ, UK

Introduction

F. Fidanza

In recent years in Western Europe lifestyles have changed remarkably. Various factors have contributed to this, but industrialization is a feature of the phenomenon. Food habits have been notably influenced; the consumer has found around him more and more varieties of foods and ways of eating, which have kept pace with technological evolution and which are also due to direct or indirect exchange with different cultures.

The public is now highly sensitive to control of body weight and to the use of dietary products, in the hope of attaining a slim, efficient, always-young image, in harmony with the rhythm of life in present-day society.

Even during pregnancy we have observed women following an irrational diet for fear of putting on too much weight! This diet was kept up with much sacrifice, as was seen from the women's answers to survey questionnaires. A drastic reduction in pasta and bread consumption, which were among the best liked foods, was often suggested by gynaecologists, with the aim of avoiding excessive weight gain. In about 25% of the cases, weight gain during pregnancy was just 6 kg. The evaluation of the nutritional status in regard to some nutrients in a group of women at the end of pregnancy showed deficient status in folates, riboflavin, thiamin and iron in various cases and in their newborn babies too, although to a lesser degree.

Our studies on adolescent girls in Perugia also showed irrational eating habits. Their nutrition knowledge was distorted and confused. The question of nutrition was taken into consideration only superficially, with no attention being paid to the particular needs of the adolescent. Out of fear of getting fat, they followed strange diets with great sacrifice, and with little success.

On the other hand, adolescent girls from hamlets and villages in the surroundings of Perugia with a lower energy intake (arbitrarily fixed at less than 20% of recommended daily intake (RDI)) showed only slight differences in vitamin (thiamin, riboflavin and folate) and mineral (iron as ferritin) status in comparison with those who had adequate energy intake.

In our longitudinal and cross-sectional studies on non-institutionalized and institutionalized elderly people in Perugia we have found a rather high prevalence of folate, riboflavin, thiamin, vitamin B6, iron and zinc malnutrition, using the current cut-off points. Considering separately the small number of elderly with lower energy intake (arbitrarily fixed at less than 25% of RDI), we found only slight differences in vitamin and mineral status in comparison with those with adequate energy intake.

In some of these elderly people we also assessed both the humoral and cellular immunocompetence. When they were compared to a group of apparently healthy young people, we observed an impairment of immune function, in particular cell-mediated, but also humoral and a decrease of the leucocytic phagocytic and bactericidal activity. The elderly people with lower energy intake in comparison with those with adequate energy intake showed only minimal differences for the immunological variables considered. This may be due to the fact that the energy intakes were not so low as to induce a change in the immunocompetence and that the difference in vitamin and mineral nutriture between the two groups was similarly too small. In addition, an adaptation mechanism cannot be excluded.

In this volume a preliminary question that needs an answer is: when is an intake low in energy? Put in this way, this question seems to me inappropriate. In fact we have reducing diets of about 2.8 MJ (1170 kcal) for women, 3.5 MJ (1460 kcal) for men and 4.0 MJ (1670 kcal) for youth of both sexes, that daily provide enough nutrients to cover the RDI. On the other hand, we have diets adequate in energy, but with rather low micronutrient content. In addition, energy intake varies according to physical activity, climate and/or individual circumstances.

As possible examples, very low energy diets are considered to be those containing fewer than 1.44 MJ (600 kcal)/day. In the fundamental Minnesota study on human starvation the semi-starvation diets provide about 4.0 MJ (1658 kcal) daily for adult men.

A more correct approach could probably be to express intakes in relation to nutrient density. Thus we can have an intake of lower nutrient density; but then we must find the critical point at which impairment of nutritional status starts. And for the nutritional status we have to consider not only the static indices, but particularly the functional ones, as we have stressed on various occasions. If the nutrient density approach is used for nutrient requirement, it will not be too difficult to find the critical point at which impairment of nutritional status starts.

In addition we have to take into account the nutrient bioavailability. Nutrient interactions have been shown with other nutrients and with non-nutritional compounds, particularly drugs. It is not easy to solve this problem. In fact we do not consume an individual nutrient, but a mixture of nutrients and very often a mixture of foods, usually handled at home in different ways. In this regard the individual conditions also have to be taken into consideration. This probably explains why we have not found a significant correlation between intake and the corresponding nutritional status for some nutrients.

Only a few problems have been outlined in this introduction: others will be considered in the volume.

I would like to make one final suggestion. To solve some of the problems just outlined at the population level we have two main possibilities. One, the most suitable for a nutritionist, is through nutrition education. But this intervention is rather costly and takes a long time to be effective. The second one is the production of foods with high nutrient density, giving preference to the most critical nutrients. We have some good examples from lay organizations. These examples can be followed successfully by official agencies.

Chapter 1

Effects of Lifestyle on Nutrient Requirements

L. Kohlmeier, G. B. M. Mensink and E. Hermann-Kunz

Introduction

In the light of current food availability it is justified to ask whether there is any reason to believe that lifestyle significantly affects nutritional adequacy. What evidence is there that lifestyle influences health status? Is there reason to believe that some aspects of our behaviour are expressly healthy? Which functional indicators of malnutrition exist in our societies?

Evidence that disease rates can be reduced through changes in behaviour comes from epidemiological comparisons between countries, examination of rate changes within countries and observations of rate changes in emigrant populations (Keys 1970, Kato et al. 1973, Haenzel 1982).

Breast cancer in women, which shows considerable intra- and intercountry variation, may be taken as an example. The age-standardized breast-cancer rates reveal clear trends in mortality of 55- to 64-year-old women with time, as in the case of the increasing rates in the Federal Republic of Germany (Hoffmeister and Junge 1988), despite increasingly successful surgical treatment. The incidence trends seen in a regional cancer registry (Statistisches Landesamt des Saarlands 1988) concur with national mortality trends. Widely different mortality rates between countries are currently seen in Europe (Fig. 1.1a) and a great variance in the time trends in mortality between countries is also present (Fig. 1.1b). Finally, strong correlational associations are seen within Europe between per capita saturated fat consumption, as measured through the primitive method of national food balance reports, and the concurrent national mortality rates (James 1988).

Indirect though these measures may be, they are concrete indicators of variation in age-standardized mortality from this disease. This variation has many causes, and differences in lifestyle are likely to be responsible for a large portion of the variation. Epidemiological and biochemical research can be expected to lead to clarification of the role of lifestyle in the development of breast cancer in the coming decade. Until now dietary factors have been suspected through alcohol consumption (Hiatt and Bawol 1984), dietary fat intake (Committee on Diet, Nutrition and Cancer 1982) and total caloric intake (Carroll et al. 1968).

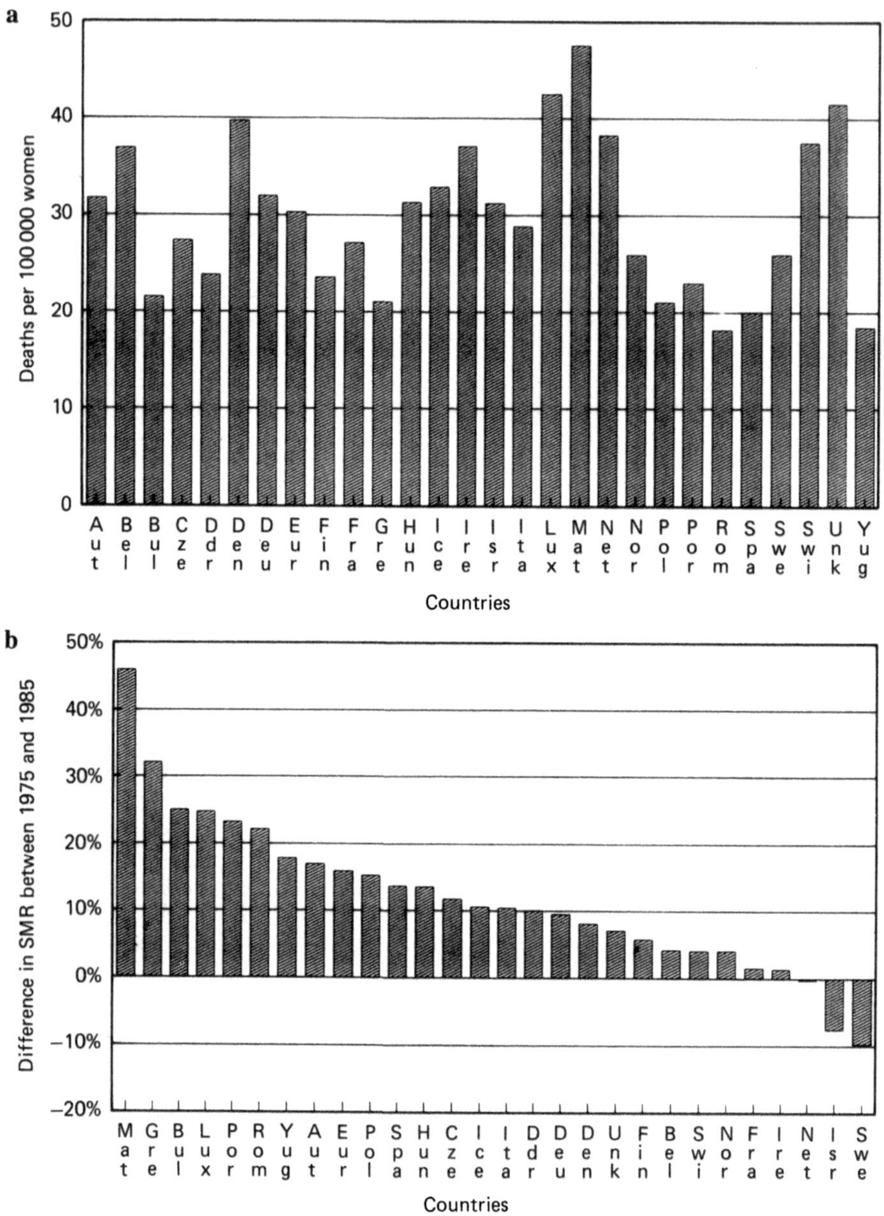

Fig. 1.1. a Female-breast cancer mortality in Europe. Standard mortality ratios, per 100 000 women, for all ages for 1985, or the most recent data available for the particular country. **b** Standard mortality ratios for malignant female-breast neoplasms, all ages, per 100 000 women; changes between 1975 and 1985.

The development of many diseases is believed to be related to dietary behaviour and the intake of specific nutrients. Such diseases include cancers of the lung, stomach, prostate, rectum, endometrium and oesophagus, cardiovascular disease, osteoporosis and osteomalacia, hypertension, anaemias, arthritis and gallstones to name the most prevalent and measurable (Henke et al. 1986,

Deutsche Gesellschaft für Ernärhung 1988, US Department of Health 1988). Changes in the rates at which these diseases occur help to determine whether a country is getting healthier or sicker. More subtle potential immunological, and neurological changes would, by their very nature, go largely undetected in the general population.

It is strongly suspected that lifestyle plays an important role in nutrient requirements. Consequently if these requirements are not reached a number of diseases can develop. This is rather amazing as such a wealth of food resources, and such a variety of intake, exists in our societies. The perspective, however, stems more from expectations of optimal health than of survival. Populations are living longer, and expect to enjoy those years without rheumatic disease, osteoporosis, neurological disorders, allergies, and a variety of largely non-fatal but disabling conditions such as gout, obesity, or adult-onset diabetes.

Some observers believe that people are consuming less food than they formerly did. Others believe that, at least partially because of increased food availability, people are eating more and getting fatter. Little information currently exists which can be used for global, European or even national analysis of changes in intakes or lifestyle. Regular monitoring systems of food consumption or health status exist in few European countries. Britain can claim a regular family food purchase survey conducted by the Ministry of Agriculture, Fisheries and Food, and the Netherlands is beginning nutritional surveillance (Löwik and Hermus 1988). National risk factors or health surveys are available, but they are not conducted in a regular or internationally standardized fashion. Information on morbidity is also widely lacking, with the exception of the cancer registries in some countries (MARC 1988). This state of affairs may change in the near future as technical resources become available, and the willingness to cooperate on a European basis increases. Otherwise it is difficult to relate differences in disease rates between countries to changes in behaviour and to differences in lifestyles. The identification of the relevant factors affecting chronic diseases remains the challenge of our decade.

Lifestyle

The World Health Organisation describes lifestyle as "a cluster of closely interrelated behaviour patterns that depend on social and economic condition, education, age and many other factors" (WHO 1988). It remains a poorly-defined entity and is readily applied as subsuming a variety of conditions. Every country and each group within a country has its own lifestyle. The difficulty lies in dissecting lifestyle successfully into measurable components which can explain the differences physiologically. In general, groups of people with a characteristic trait are compared with other "normal people". Interestingly, many of the groups formerly believed to be at a greater health risk due to their lifestyle are now believed to be holding the advantage.

Vegetarianism is one lifestyle currently in vogue. It was formerly believed that vegetarians were at a greater risk of ill health, either because of vitamin or mineral deficits, or because of the suspected greater risk of cancer from higher nitrate intakes. Recent studies suggest that the overall and relative risk of cancer

Table 1.1. Mortality ratios for selected cancer sites among white Californian Seventh Day Adventist (1960–76) and non-Seventh Day Adventist (1960–71) participants in the American Cancer Society study and all US whites 35 years of age and older (adapted from Philips et al. 1980)

ICD code 8	Primary cause of death	Sex	Age–sex-adjusted mortality ratio	
			SDA: non-SDA	SDA: US whites for 1969–75
140–209	All cancer	m	0.60**	0.52
		f	0.76**	0.68**
151	Stomach cancer	m	1.41	0.93
		f	0.89	0.62**
153–154	Colo-rectal cancer	m	0.62**	0.57**
		f	0.58**	0.51**
162	Lung cancer	m	0.18**	0.17**
		f	0.31**	0.34**
	Other smoking[a] related cancers	m	0.59**	0.40**
		f	0.74**	0.59**
174	Breast cancer	f	0.85	0.90
185	Prostate cancer	m	0.92	0.87
	All other cancers[b]	m	0.74*	0.52**
		f	0.89	0.70**
001–999	All causes	m	0.66	0.51
		f	0.88	0.60

*, **: $p<0.01$ and $p<0.05$ respectively, based on chi-square.
[a] This group included mouth and pharynx (140–149), oesophagus (150), larynx (161), bladder and other urinary organs (188, 189.2–189.9) and pancreas (157).
[b] This group included all cancer sites not included in the specific categories above.

of vegetarians is lower, as seen in Table 1.1, where Seventh Day Adventists are compared with non-Seventh Day Adventists. Some of the results may be related to specific factors which characterize this close religious community.

There are few studies of European vegetarians; a German group of about 370 vegetarians and a control group of health-conscious non-vegetarians comparable in age, sex and school education level, were investigated (Rottka et al. 1988). The studied group was made up of 72% of ovo-lacto-vegetarians and 25% lacto-vegetarians. Few subjects were strict vegans (3%), subjects who avoid all animal foods, including even honey.

This vegetarian diet, as measured by 7-day dietary diaries of females, is in contrast to the non-vegetarian nutrition, characterized by a generally lower energy intake (Fig. 1.2a), and a relatively similar intake of fat (Fig. 1.2b).

Despite the lower energy intakes of the vegetarians, great differences between mean daily mineral and vitamin intake were seen (Fig. 1.2c, d). Except for vitamin D, niacin and vitamin B12, vegetarians have a higher intake than the controls. The vegetarian diets were richer in minerals, particularly phosphorus, calcium and magnesium. The tables give the findings for females, but the results were the same for the diets of male vegetarians.

Undesirable differences in nutrient levels were only found in this German vegetarian population in relation to the iron status in some of the long-term female vegetarians, where iron-deficiency anaemias were measured. Serum B12

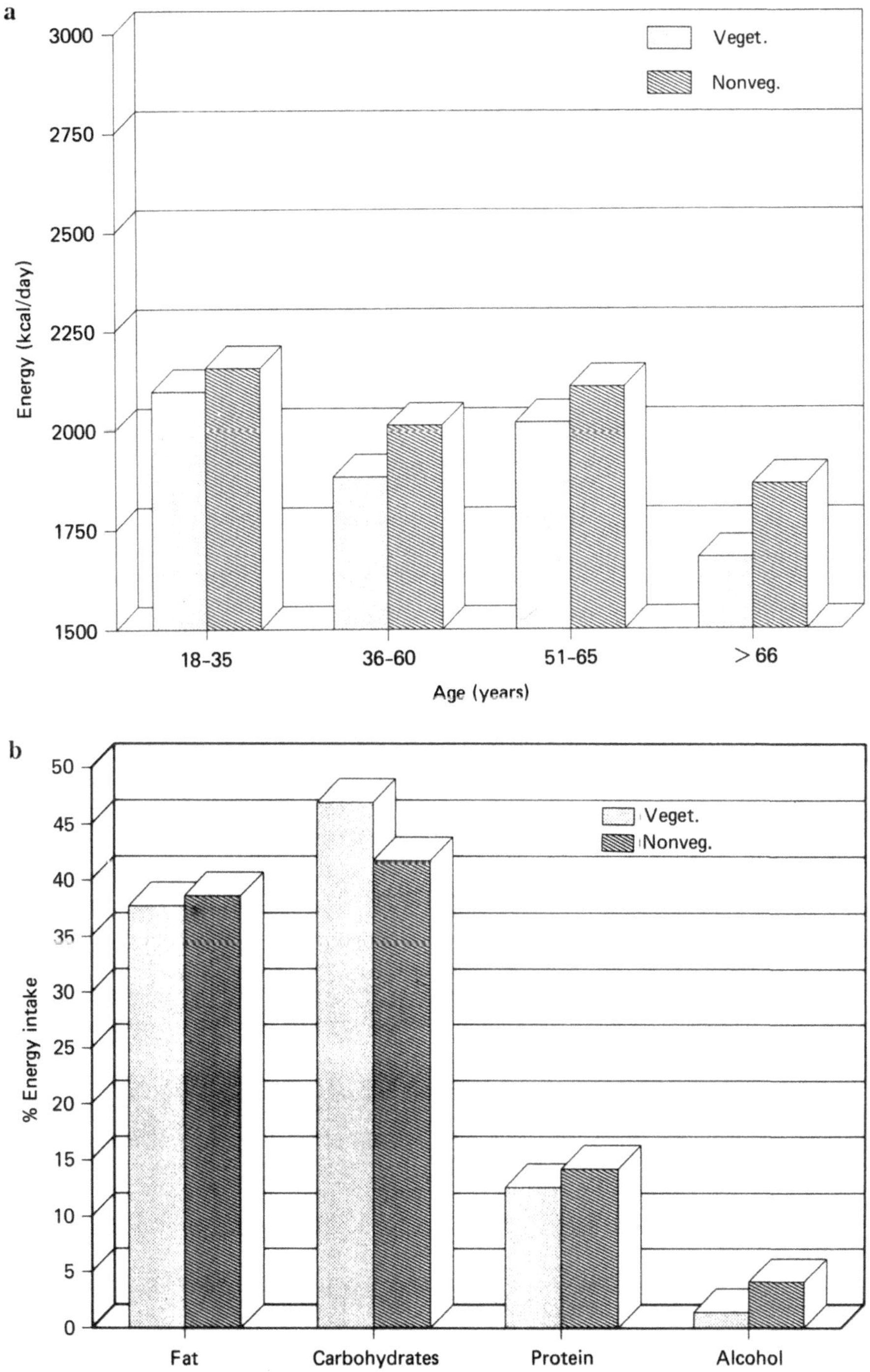

Fig. 1.2. **a** Mean energy intakes of female vegetarians and non-vegetarians. **b** Macronutrients as mean percentage of energy intake for female vegetarians and non-vegetarians.

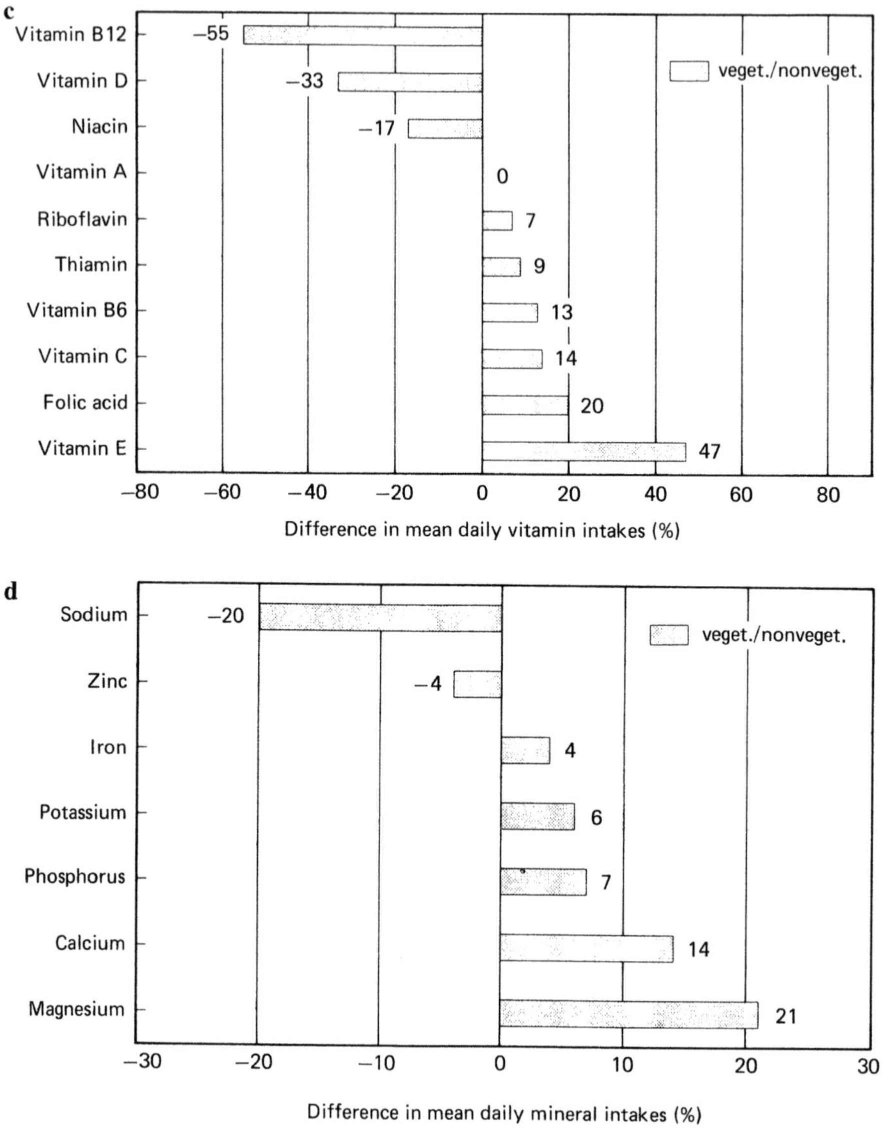

Fig. 1.2. c Percent difference in mean daily vitamin intake between female vegetarians and non-vegetarians. **d** Percent difference in mean daily mineral intake between female vegetarians and non-vegetarians.

is also lower, but this does not pose an increased risk of macrocytosis (pernicious anaemia).

Lower risk levels of cardiovascular disease in general were found in these vegetarians. Analysis of the medical data shows that despite their high fat intakes vegetarians have lower blood pressure, lower total serum cholesterol, lower serum triglycerides and also a lower body mass index than the controls (Deutsche Gesellschaft für Ernährung 1988).

The findings in vegetarians present a good example of the problem of defining lifestyle as an entity with clearly defined components. In order to be scientifically useful, however, these differences must be related to individual aspects of vegetarian lifestyle. If we cannot do this we will remain unable to extrapolate the consequences of individual risk-related decisions in daily life. Without this approach, one would have to consider all aspects of Seventh Day Adventists' lifestyles, including church going, tax paying, home building, children raising, etc. rather than separating out particular characteristics in order to explain their low incidence of cardiovascular diseases and cancers (Phillips et al. 1980).

Lifestyle Components

The basic model underlying our studies of diet and health has a black box nature. One is looking for effects of food intake on the incidence or risk of disease without exact knowledge of exactly what is taking place at the molecular and cellular level.

In relation to this model the question of lifestyle and nutrient interactions has two components, "Does lifestyle affect requirements for or availability of consumed nutrients directly?" and "Does lifestyle affect food intake?" These questions are expressed diagrammatically in Fig.1.3, b).

Single components which can affect nutrient requirements and disease onset include tobacco use, alcohol consumption, physical activity, use of medications, and food habits. Their potential effects on nutrient status are outlined below.

Smoking

Cigarette consumption is high in Europe. A large proportion of the whole population in every country is smoking regularly. The annual consumption in cigarettes per person for 1983–1986 is shown in Fig. 1.4 (WHO 1989). Projections for the year 2000 predict increases in cigarette consumption in a majority of European countries. The WHO goal of health for all by the year 2000, that the amount of tobacco consumed be reduced by 50%, with a minimum of 80% of the population non-smokers, may not be realized. Smoking is currently a powerful factor in the lifestyle in Europe. In the Federal Republic of Germany, one-third of the population smokes, and more girls and young women are smoking than ever before (Bayerisches 1986). This trend is characteristic of a number of countries, and presents unfavourable trends in an especially vulnerable population in terms of nutritional effects. This is the group which should be optimally nourished before pregnancy, and the group which needs to prepare itself nutritionally for a life expectancy of some 80 years. It is known that smoking women have a greater risk of osteoporosis, which is hypothesized to be due to lower oestrogen production among smokers (US Department of Health 1988). Other interactions include effects on ascorbic acid metabolism and an increased basal metabolic rate.

Numerous differences between circulating vitamin levels in smoking and non-smoking women were found in our studies (Arab et al. 1982). The differences in mean levels of circulating vitamins are presented in Fig. 1.5. Smokers have

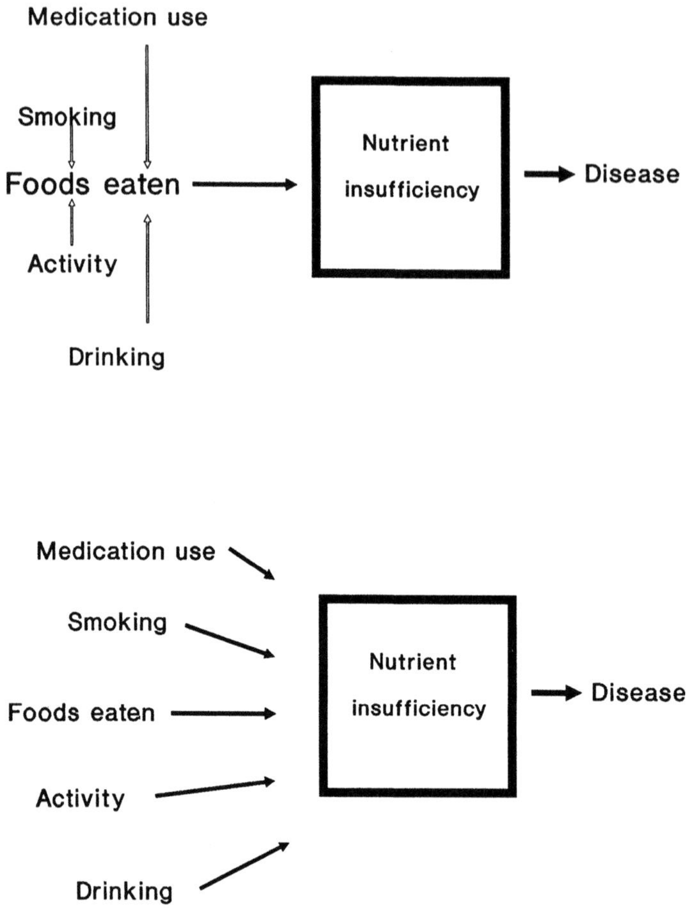

Fig. 1.3. **a** Direct effect of lifestyle on nutrient insufficiency. **b** Indirect effects of lifestyle on nutrient insufficiency.

greatly reduced levels of folate and vitamin B12 in serum (24% and 13% respectively). Beta-carotene levels are 11% lower in serum, vitamin C is 6.8% lower and the vitamins D and E showed 4% lower levels.

All told, smoking is a component of European lifestyle which affects nutrient availability at two levels, through altered food selection, and by influencing nutrient metabolism and therefore requirements.

Alcohol

The potential effects of ingestion of alcohol on requirements and intake, digestion, absorption, transport, storage, metabolism and excretion of many other nutrients are illustrated in Fig. 1.6. Complex interactions underlie these documented effects, which are related to the type, quantity, frequency and duration of alcohol consumption and are intertwined with the food and nutrient

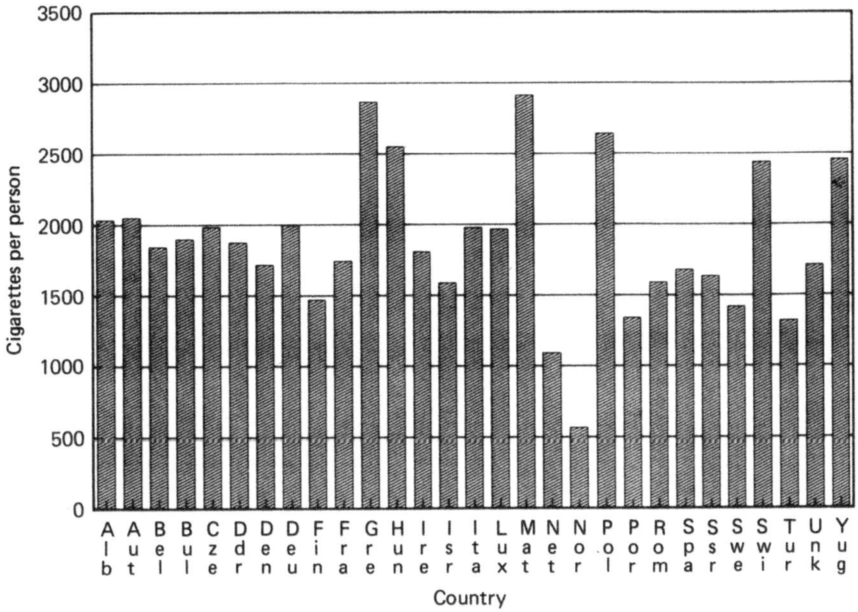

Fig. 1.4. Annual cigarette consumption for countries in Europe.

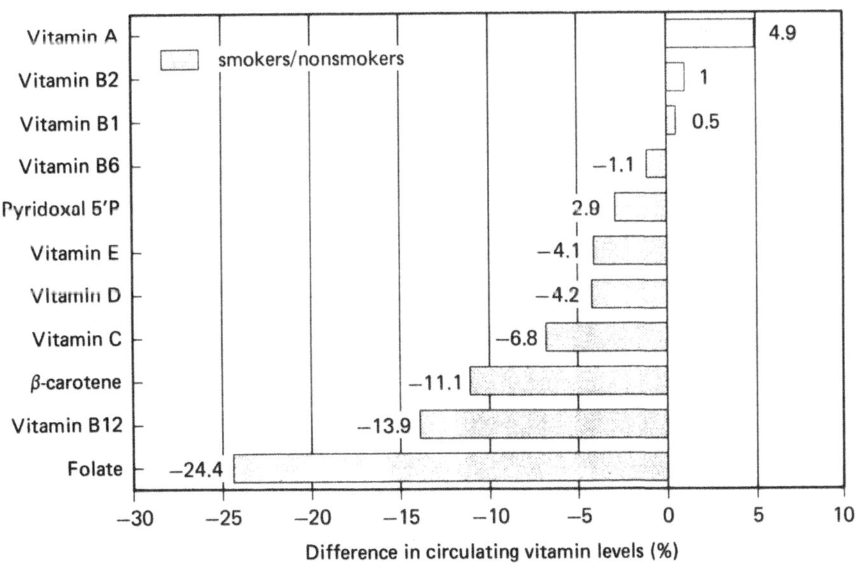

Fig. 1.5. Percent difference in circulating vitamin levels between 20–40 year-old female smokers and non-smokers.

intake of the individual. Little is known of the consequences other than the occurrence of liver disease. The international mortality rate comparisons of liver cirrhosis reveal the expected rate differences between countries. Intra-national studies reveal clear dose-response relationships between reported alcohol intakes and the levels of liver enzymes in serum in the normal population (Arab et al. 1982).

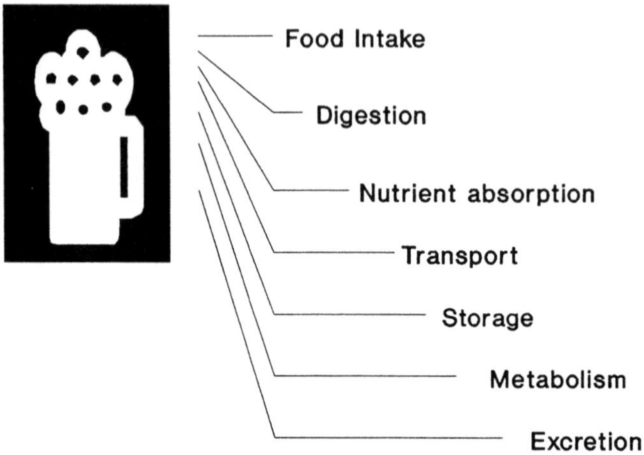

Fig. 1.6. Influences of alcohol on nutrient adequacy.

The health consequences of high alcohol consumption are not only in the area of liver cirrhosis and alcoholism. Traffic and other accidents, suicides, homicides, birth defects, and the risks of cancer of the oesophagus, liver, mouth and other sites are increased with regular alcohol intakes.

Nutritional deficiencies have long been recognized in alcoholics, and the beneficial effects of supplementation of protein and B vitamins was recognized over 40 years ago (Patek and Post 1941). Regular drinking can result in pancreatitis and pancreatic insufficiency, impaired secretion of bile salts, and abnormalities in intestinal mucosal cells, each of which will have consequences on fat digestion. The effects can be seen in the availability of fat-soluble vitamins. In the case of vitamin A an accompanying reduction in the synthesis of retinol-binding protein is also seen. Vitamin D levels can also be affected. Higher than expected rates of osteomalacia and osteoporosis are seen amongst alcoholics (US Department of Health 1988). Effects on levels of vitamin E are theoretically plausible but clinical evidence of deficiencies does not exist. On the other hand, vitamin K deficiency due to fat malabsorption can occur and this results in blood clotting disorders in alcoholics.

A number of other vitamins and minerals are affected by alcohol consumption. Thiamin absorption and activity can be reduced, and its storage inhibited by fatty infiltration of the liver. Wernicke–Korsakoff syndrome in alcoholics is reversible through thiamin supplementation. Effects on riboflavin and niacin availability have not been proven. The absorption of vitamin B6, its release from the liver and metabolism are affected by alcohol. Folate deficiency is one of the most common vitamin deficiencies in alcoholics, probably due to damage to the intestinal mucosa. Alcohol increases urinary excretion, and inhibits the enzymes involved in folate metabolism.

The potential effects of alcohol on nutrient availability do not stop here; the minerals known to be vulnerable to alcohol consumption include iron, calcium, magnesium and zinc. The iron status can be affected in either extreme deficiency due to losses from gastrointestinal bleeding, or iron overload due to increased intake and absorption (through increased ferric iron solubility). Calcium requirements are affected by alcohol consumption as the urinary excretion increases, as is the case with zinc and magnesium.

Thus, alcohol consumption is one of the most important factors related to lifestyle and nutrient interactions in our societies, both in its effects on eating behaviour (Polivy and Herman 1976) and on metabolism.

Physical Activity

Our societies are, as has been lamented, extremely inactive. The major energy expeditures in the majority of our populations are due to basal metabolic rates, resting, sitting and sleeping, of which we do so much and, therefore, little else. In our studies of elderly persons, sleep accounted for the greatest energy expenditure (27%), sitting and standing for another 30%. Of their average daily expenditure of 36.2 kcal per kilogram body weight, only 16 kcal were from "activities" such as walking, light work or sports. The relationships can be seen in Fig. 1.7.

Studies of the effects of inactivity and of heightened activity levels are just beginning, partly due to the findings of reduced cardiovascular risk, and improved tertiary prevention through activity programmes. Also, the risk of cancers in women have been related to their physical activity levels (Frisch 1985). But basically there is no direct information on the absolute energy expenditure levels of populations, and consequently no information on trends. Whether energy expenditure is increasing, decreasing or has been stable for decades is unknown.

Assessment of physical activity in representative surveys is almost as difficult as assessment of food intake. The results presented here are based on 7-day activity diaries in which subjects record all time spent, in hours and minutes throughout the day. Despite the heavy burden on the respondents with this

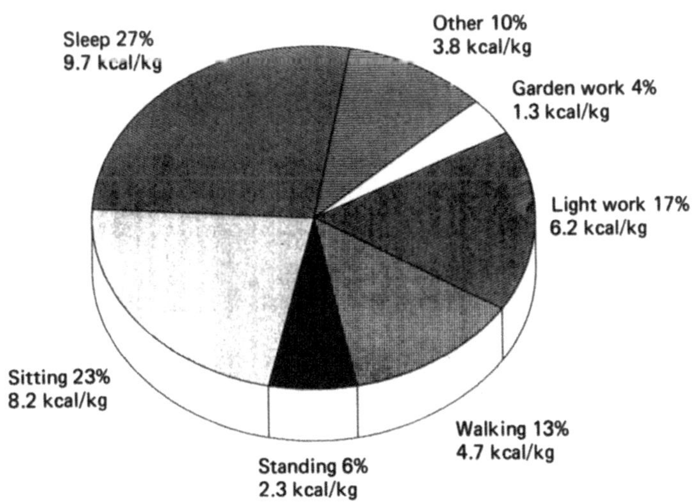

Fig. 1.7. Average daily energy expenditure involved in different activities in older persons.

method, 77% of the senior (n=385) and 84% of the junior (n=395) population filled in and returned complete, usable records. The amounts of time spent in different activities was converted into estimates of energy expenditure per activity which were summarized into daily totals. The basis of this calculation was the MET unit. The MET-level for a certain activity is measured as the oxygen consumption per kg body weight. One MET describes the oxygen consumption at rest of 3.5 ml/min. This equals the basal metabolic rate per kg body weight of 1 kcal per hour (1 MET=1 kcal/kg × h).

A MET level of 24 per day would be the theoretical minimum, and would mean the individual sleeps 24 h a day. A person who sleeps 8 h and sits 16 h a day would already have a daily MET-level of 32. For a person of 70 kg body weight this would mean an energy expenditure of 70×32=2240 kcal/day.

A closer look at the distribution of energy expenditure levels in the above-mentioned elderly reveals that the activity levels of very few persons are so great that an effect on nutritional requirements due to increased energy expenditure needs can be expected (Fig. 1.8). The median and mode intakes are approximately 38 kcal energy expenditure per kilogram body weight. The population is quite passive. Were this to change in the future, physical activity in any regular fashion affects the total energy expenditure so strongly that clear differences in nutrient intake and needs between physically active and inactive people could be expected.

In this population, comparisons of active and inactive elderly show numerous differences in intake and nutritional status (Mensink and Arab 1989). Comparison of the body mass index distributions of active and inactive older men showed clear differences between these two groups (Fig. 1.9). Fewer active men were over or underweight, as compared to the inactive. Lower weights among the inactive were partially due to illness. Remembering that muscle mass is not accounted for in these measurements, clearly the group of active men have a more ideal weight distribution than the inactive.

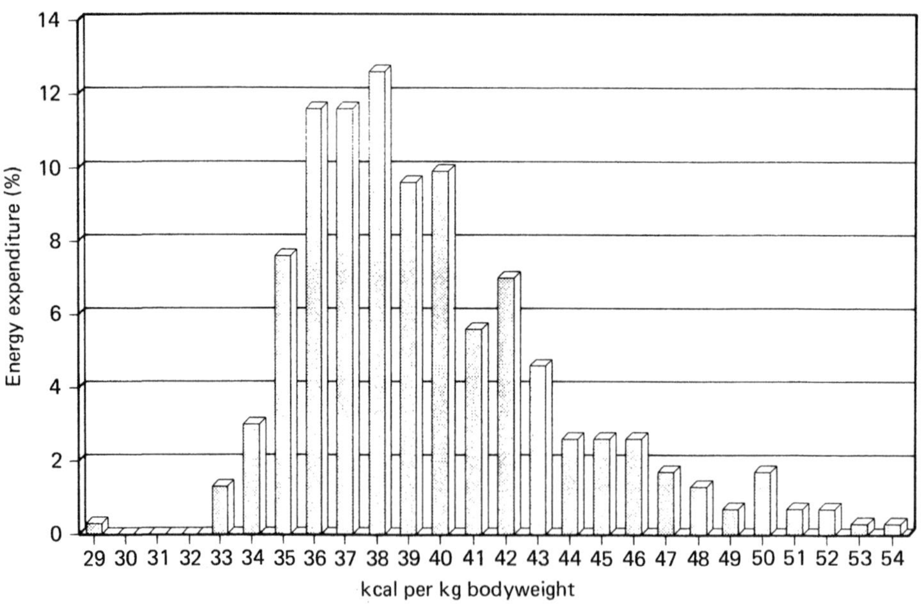

Fig. 1.8. Energy expenditure distribution for men aged 65 to 75 years.

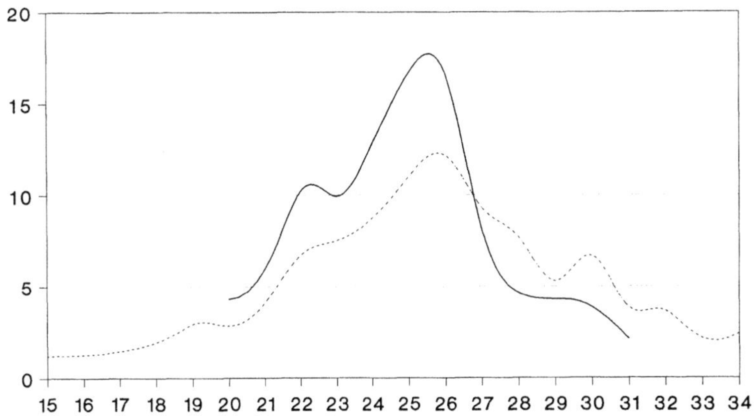

Fig. 1.9. Body mass index distribution of active (full line) and inactive men (dashed line).

Comparisons of these higher and lower activity groups of the elderly men showed that the total intakes of a number of nutrients were greater among more active elderly men (Fig. 1.10a). This included vitamin A and beta-carotene and vitamin E (20% difference between group means), vitamin C and calcium (17%, 15%), iodine, riboflavin and vitamin B12 (7%–10%). In some cases this was due to a 7% greater caloric consumption. However, the nutrient density was also higher for vitamins A, E and C and for calcium (Fig. 1.10b). Surprisingly, the thiamin intakes, both absolute and relative, were slightly lower in active elderly men than in the inactive.

Different levels of circulating nutrients were found in both directions. Particularly in the case of vitamin D, higher serum levels were noted which were not directly attributable to differences in food intake between the groups (Fig. 1.10c). The carotenoids, as expected due to the higher intakes, were higher among the active. Tocopherol levels, on the other hand, were distinctly lower. Marginal differences in vitamin A, thiamin, and riboflavin were noted in a direction unfavourable to the active.

Medication

Regular use of medication has become a part of our lifestyle. In the national health survey of the Federal Republic of Germany of 1985–86 it was seen that 60% of the 25–69 year-old population reported regular use of medication (Melchert et al. 1987). The elderly particularly represent a group at high risk of nutrient insufficiency due to medication use. Not only is the use of multiple medications common among the elderly, with 50% of this group taking three or more drugs concurrently (Arab 1985), but it is reported that they may show reduced metabolic abilities, so that the effects are longer lasting.

Similarly to the potential effects of alcohol on nutrient availability, different drugs can affect appetite, digestion, absorption, and the metabolism of nutrients and their utilization and excretion. Appetite can be affected in both directions, either stimulated or reduced by specific drugs. Digestion and absorption of nutrients can be affected by laxatives which stimulate transit time, antacids,

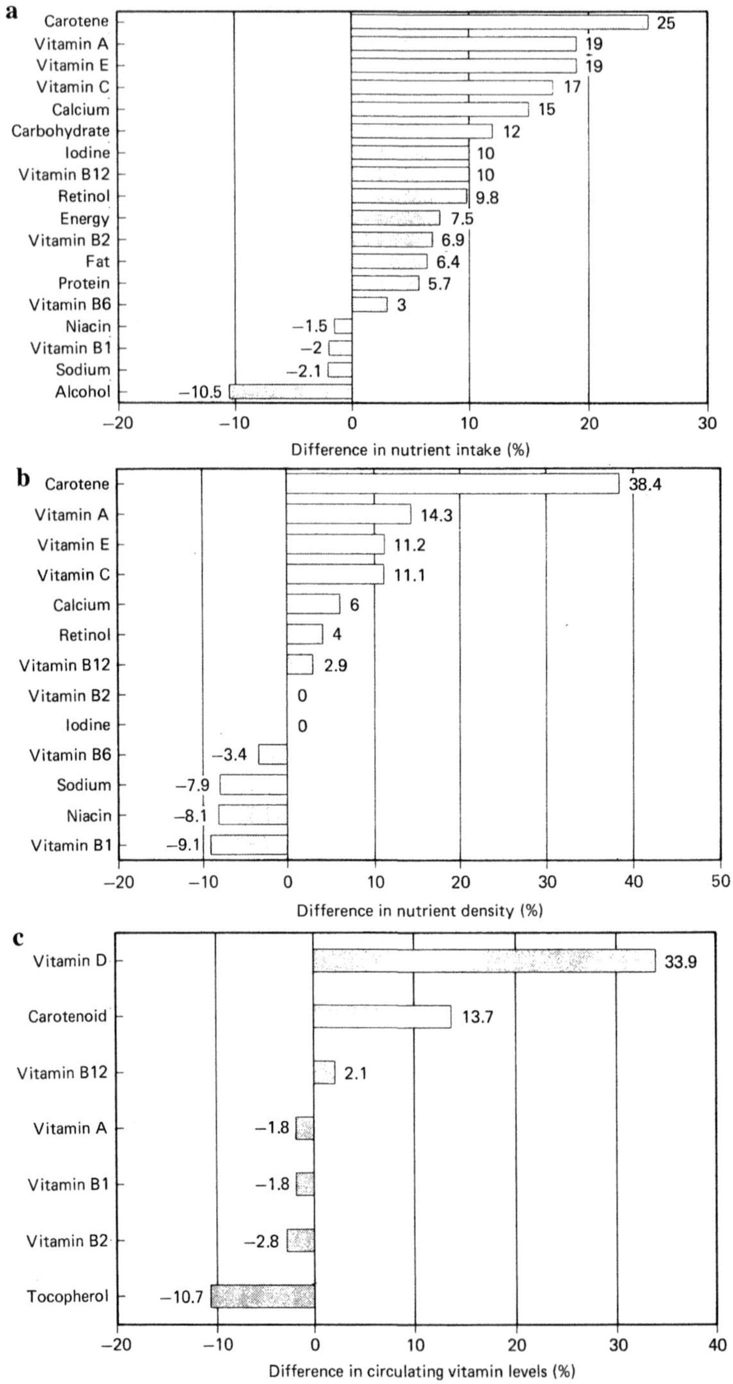

Fig. 1.10. a Percent difference in nutrient intake per day between active and inactive men aged 65 to 74 years. **b** Percent difference in nutrient density in the food between active and inactive men aged 65 to 74 years. **c** Percent difference in circulating vitamin levels between active and inactive men aged 65 to 74 years. In all parts of this figure the stippled areas are the active : inactive ratio.

which can affect phosphate, copper or calcium absorption, bile acid sequestrants which can bind fat-soluble vitamins, and antibiotics, some of which can affect the absorption of vitamin K and others the absorption of retinol. The birth control pill, because of its regular and widespread use, has been the particular subject of studies of nutrient effects. Some evidence of reduced utilization of B vitamin in pill users has been described. The class of anticonvulsants can affect folate, vitamin D or thiamin status. Barbiturates may reduce metabolism of vitamin K. Particularly through the regular use of diuretics, excretion of minerals becomes an issue of concern (US Department of Health 1988).

Clearly, many potential effects are present in the area of nutrient status and medication use. The active ingredients are so varied, the intra-individual response so differing, and the preliminary nutrient status of such importance that the science of pharmacoepidemiology has a large task ahead. When the synergistic and inhibitory effects of multiple medication use are considered, the problems seem almost endless. Lastly, the consumption of pharmacological doses of vitamins and minerals also affects the nutrient status, of both the nutrient itself, and of its competitors. This is a lifestyle which seems to be increasing in prevalence, in the hope of "supernutrition" through excessively high self-dosing.

The Epidemiological Model

The components of lifestyle as they may affect nutrient adequacy have been examined individually in this review. This differs from the approach which the epidemiologist normally applies in the analysis of the effects of lifestyle. The effects of different "exposures" on a single outcome should be examined simultaneously, and quantified in relation to one another. In the current example, the influences of all possible lifestyle effects on a single parameter of nutrient sufficiency would be examined. This allows the identification of major determinants of nutrient adequacy or inadequacy in a given population. An example of this in relation to the thiamin sufficiency of the German population will be used as an illustration (Fig. 1.11).

Fig. 1.11 shows a model which predicts the thiamin status and tries to explain the high levels of high activation potential among the population as a function of their intakes, smoking behaviours, alcohol consumption, use of medications, levels of body fat and physical activity, age and sex, as well as seasonal effects.

Our findings, which certainly apply for the German population, show that alcohol consumption and smoking behaviour are the major determinants of thiamin status. The general intakes are sufficient to cover those recommended but, taking the level of metabolic abnormality as criteria for undersupply, metabolism is not adequate for the lifestyles of up to one third of the adult population. Examination of the effects of cooking, of regular use of thiaminases, for example in coffee or tea consumption, and of oral contraceptives revealed that these are not primary determinants of thiamin insufficiency. They each play an insignificant role in the elevated enzyme activation levels. In relation to food intake, a greater energy intake was associated with higher levels, as long as the energy did not come from higher fat intakes (Arab-Kohlmeier et al. 1988).

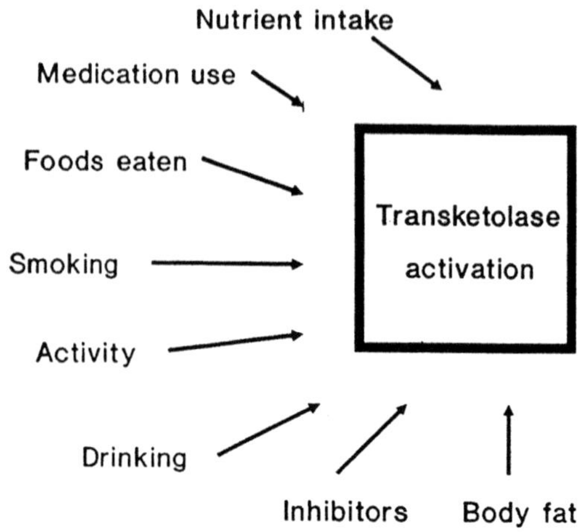

Fig. 1.11 Eiidemiological model of thiamin sufficiency.

Conclusion

It can be said our lifestyles can and do play an important role in nutrient sufficiency. In many cases, in modern European countries, lifestyle is more important than the absolute intake of foods and nutrients in determining the nutrient levels available to the body. Modern lifestyle are not stagnant, they are changing drastically, but the exact magnitude of the changes are unknown. The changes are bound to affect nutrient status and the health of society. Nutritional monitoring systems in Europe are needed which include indicators of bodyweight, activity levels, smoking, drinking, and medication use. This would allow the detection of detrimental trends and reorientation of priorities and recommended levels to suit the lifestyles of the population.

References

Arab L, Schellenberg B, Schlierf G (1982) Nutrition and health: a survey of young men and women in Heidelberg. Ann Nutr Metab 26:1–244

Arab L (1985) Ernährungszustand von Senioren – Ergebnisse einer repräsentativen Bevölkerungsuntersuchung. Ernährungs-Umschau 32:67–71

Arab-Kohlmeier L, Sichert-Oevermann W, Schettler G (1988) Thiaminzufuhr und Thiaminstatus der Bevölkerung in der Bundesrepublik Deutschland. Springer-Verlag, Berlin

Bayerisches Landesministerium des Innern, Bayerisches Staatsministerium für Arbeit und Sozialordnung (1986) Alkohol Drogen Medikamente Tabak: Jugend fragt Jugend, Präsentativerhebungen bei Jugendlichen in Bayern 1973, 1976, 1980, 1984

Carroll KK, Gammal EB, Plunket ER (1968) Dietary fat and mammary cancer. can Med Assoc J 98:590–594

Committee on Diet, Nutrition and Cancer, National Research Council (1982) Diet, nutrition, and cancer. National Academy Press, Washington DC

Deutsche Gesellschaft für Ernährung (1988) Ernährungbericht 1988. Druckerei Henrich, Frankfurt am Main

Frisch R (1985) Lower prevalence of breast cancer and cancers of the reproductive system among former college athletes compared to non-athletes. Br J Cancer 52:885–891

Haenzel W (1982) Migrant studies. In: Schottenfeld D, Fraumeni JF (eds) Cancer epidemiology and prevention. Saunders, Philadelphia, pp 194–207

Henke KD, Behrens C, Arab L, Schlierf G (1986) Die Kosten ernährungsbedingter Krankheiten. Band 179 der Schriftenreihe des Bundesministers für Jugend, Familie, Frauen und Gesundheit. Kohlhammer, Stuttgart

Hiatt RA, Bawol RD (1984) Alcoholic beverage consumption and breast cancer incidence. Am J Epidemiol 120:676–83

Hoffmeister H, Junge B (1988) Mortalität an Ernährungsbedingte Krankheiten. Deutsche Gesellschaft für Ernährung, Ernährungsbericht 1988:36–41

Hoffmeister H, Stolzenberg H, Schön D, Thefeld W, Hoeltz J, Schröder E (1988) Nationialer Untersuchungs-Survey und regionale Untersuchungs-Surveys der DHP Band II DHP Forum, Deutsche Herz-Kreislauf-Präventionsstudie

James WPT (1988) Healthy nutrition: preventing nutrition-related diseases in Europe. WHO Regional Publications, European Series No. 24, Copenhagen

Kato H, Tillotson J, Nichaman MZ, Rhoades GG, Hamilton HC (1973) Epidemiologic studies of coronary heart disease and stroke in Japanese men living in Japan, Hawaii, and California. Serum lipids and diet. Am J Epidemiol 97:372–85

Keys A (1970) Coronary heart disease in seven countries. Circulation 41 [Suppl 1]

Löwik MRH, Hermus RJJ (1988) The Dutch nutrition surveillance system. Food Policy 359–365

MARC (1988) Health monitoring in the prevention of disease caused by environmental factors. Report of a WHO/CEC Workshop

Melchert H-U, Braemer-Hauth M, Gerding B (1988) Arzneimittelkonsum in der Bundesrepublik Deutschland. Tätigkeitsbericht 1987 Bundesgesundheitsamt; 154–155. MMV Medizin Verlag, Munich

Mensink GBM, Arab L (1989) Relationships between nutrient intake, nutritional status and activity levels in an elderly and in a younger population; a comparison of physically more active and more inactive people. Z Gerontol 22:16–25

Patek AJ Jr, Post J (1941) Treatment of cirrhosis of the liver by a nutritious diet and supplements rich in vitamin B complexes. J Clin Invest 20:481–505

Phillips RL, Garfinkel L, Kuzma JW, Beeson L, Lotz T, Brin B (1980) Mortality among California Seventh-day Adventists for selected cancer sites. J Natl Cancer Inst 65:1097–1107

Polivy J, Herman CP (1976) Effects of alcohol on eating behavior: disinhibition or sedation? Addict Behav 1:121–125

Rottka H, Hermann-Kunz E, Hahn B, Lang H-P (1988) Deckung des Nährstoff- und Nahrungsenergiebedarfs bei Vegetariern. Deutsche Gesellschaft für Ernährung, Ernährungsbericht 1988:284–306

Statistisches Landesamt des Saarlands (1988) Morbidität und Mortalität an bösartigen Neubildungen im Saarland 1986, Sonderhefte des Statistischen Amtes des Saarlandes (Saarbrücken) 143–88

US Department of Health and Human Services (1988) The Surgeon General's report on Nutrition and Health. DHHS (PHS) Publication no. 88–50210. US Government Printing Office, Washington, DC

World Health Organization, Regional Office for Europe (1988) Research Policies for Health for All. WHO Copenhagen World Health Organization, Regional Office for Europe (1989). Health for All Computer Program. WHO, Copenhagen

Chapter 2

Effects of Special Physiological Conditions on Micronutrient Status: Pregnancy, Lactation and Ageing

H. van den Berg

Introduction

Nutritional status can be considered to reflect the amount of (micro) nutrient available in the body to enable biochemical (physiological) processes that depend on that specific nutrient for their function. Actually, the nutritional status reflects the balance between nutrient intake and nutrient requirement. Both nutrient requirement and intake are variables depending on various factors, such as body build and composition. For many micronutrients needs are associated with energy and/or macronutrient intake. Among the determinants are the metabolic state as well as the physiological condition of the individual. During growth, but also in pregnancy, metabolism is strongly anabolic, while in the last phase of life the catabolic state may be more dominant, as a result of (chronic) disease. Between these two extremes, metabolism is mainly in the steady state, at least when no disease is present. In this state metabolic control is directed to maintain the constancy of the internal milieu, a situation or condition referred to as homeostasis.

A remarkable and interesting phenomenon or quality of the (human) organism is its ability to (temporarily) change this balance or "setpoint of metabolic control" when entering a "new" physiological state. Kennedy (1967) introduced the term "homeorhesis" to describe the "orchestrated change for the priorities of a physiological state" or the coordination of metabolism in various organs to support a physiological state. Such a change in metabolic control may have consequences with respect to (micro) nutrient partitioning, i.e., nutrient absorption and retention, as well as distribution between tissues and circulation. Bauman and Currie (1980) further elaborated this concept for the situation during pregnancy and lactation. During pregnancy the fetoplacental unit competes with other organs and tissues for nutrients. The placenta is very effective in accumulating nutrients and passing them to the fetus. The supply of nutrients to the fetus seems therefore well protected, although the early phase of pregnancy, when the placenta has not yet fully developed, may be critical. The fetus behaves like an intelligent parasite, able to achieve subtle changes in maternal physiology, probably via an effect on the hormonal balance.

The ageing process can also be considered as a changing metabolic state due to age-induced changes in organ function, and a progressive loss of homeostatic control. Knowledge of the "adaptational" changes in nutrient partitioning and metabolism is essential for proper assessment of the nutritional status, i.e., for selection and interpretation of parameters to be used in nutritional status assessment.

"Physiological condition" is a rather vast concept; and only some of the general principles and aspects of a different metabolic control or physiological condition on the micronutrient status will be discussed, with special emphasis on the effects of pregnancy, lactation (temporary increase in nutrient need; homeorhetic and homeostatic control) and ageing (decreased intake; impaired homeostatic control). More specific information with respect to specific micronutrients is presented elsewhere.

Effect of Pregnancy and Lactation on Micronutrient Status

The Nutrient Cost of Pregnancy and Lactation

There can be no doubt that nutrient requirements increase during pregnancy as a consequence of the increasing fetal demands and the enlargement of maternal reproductive organs such as breast and placental tissue. During lactation "extra" nutrients are needed for the production of breast milk. These "additional" requirements may, at least in pregnancy, be partly covered by an increase in maternal metabolic efficiency, or may be derived from pre-existing maternal stores, but they have to be covered largely by an increase in dietary intake. The relative increments in the NAS/NRC (9th edn., 1980) recommendations for pregnant and lactating women are summarized in Table 2.1, and are based upon estimates of the "nutrient cost" of pregnancy and lactation. The considerable diversity in recommendations from various Nutrition Councils and authorities is to be attributed to the problems and uncertainties in estimating the nutrient cost. The question "how much extra is needed" has been approached at various levels and with various methods, such as the factorial and epidemiological approach (Beaton 1979; Sandstead 1981).

The nutrient cost of lactation is probably similar to the amount lost via daily milk secretion, taking into account absorption rate and bioavailability of the nutrients from the diet. As illustrated in Table 2.1, recommendations for lactating women are generally higher than for pregnant women. Approximate data for daily secretion in milk of some vitamins, minerals and trace elements are summarized in Table 2.2. The breast milk content of most vitamins, but probably not of trace elements, depends on the dietary intake and nutritional status of the mother (Lönnerdal 1986). For some micronutrients the length of lactation also affects the nutrient content in the milk (Karra et al. 1986). When we compare the approximate nutrient content in the milk of well-nourished mothers with the recommendations in Table 2.1, the recommended daily amounts (RDAs) seem

Table 2.1. Recommended dietary allowances (RDA) for pregnant and lactating women as compared to those for adult non-pregnant women (calculated from the 9th edition of the NAS/NRC (1980)

Nutrient	Non-pregnant	Relative increase (%)	
		Pregnancy	Lactation
Energy (MJ)	8.4	15	25
Protein (g)	44	68	45
Vitamin A (μg RE[a])	800	25	50
Vitamin D (μg)	5	100	100
Vitamin E (mg ∝-TC[b])	8	25	37
Thiamin (mg)	1.0	40	50
Riboflavin (mg)	1.2	25	42
Vitamin B6 (mg)	2.0	30	25
Folate (μg)	400	100	25
Vitamin B12 (μg)	3.0	33	33
Vitamin C (mg)	60	33	66
Calcium (mg)	800	50	50
Magnesium (mg)	300	50	50
Zinc (mg)	15	33	66
Iron (mg)	18	(166)	(166)

[a] RE, retinol equivalents.
[b] alpha-TC, alpha-tocopherol.

Table 2.2. Approximate micronutrient contents in mature breast milk from well-nourished mothers (DHSS 1977)

Nutrient	Content	Nutrient	Content
Vitamin A (μg/l)	400	Vitamin C (mg/l)	50
Vitamin E (mg/l)	3		
Vitamin D (μg/l)	0.05	Calcium (mg/l)	350
Thiamin (μg/l)	200	Magnesium (mg/l)	30
Riboflavin (μg/l)	450	Iron (mg/l)	0.7
Vitamin B6 (μg/l)	100	Zinc (mg/l)	3
Vitamin B12 (μg/l)	0.5	Copper (mg/l)	0.4
Folate (μg/l)	40	Selenium (μg/l)	20

generous for some nutrients, such as vitamin B6, magnesium and zinc.

Estimation of the nutrient cost of pregnancy is more complex and beset with more uncertainties. A general problem in studies on nutrient requirements is that every clinical or biochemical criterion used for assessment has its "own" requirement (Beaton 1986). In pregnancy, this problem is even more complicated as "physiological adjustments", directly or secondarily, affect nutrient metabolism and distribution (*see* next section).

This can be illustrated by the different estimates of the vitamin B6 requirement of pregnant women using various criteria and/or parameters (Table 2.3). Following a factorial approach the vitamin B6 content of the conceptus can be estimated to amount to 10–20 mg, assuming a mean lean tissue vitamin B6 content of ca. 5 μg/g and a total lean body mass of about 50%. Assuming that this amount is transferred to the fetus in the last trimester, the "additional" vitamin B6 requirement is 0.2 mg/day. Using "biochemical normality" as the criterion, i.e., estimating the amount required to keep biochemical parameters within the (non-pregnant) reference range, results in much higher estimates. Roepke and Kirksey (1979) found an association between an inadequate vitamin B6 status and unsatisfactory Apgar scores for infants at birth (after 1 min). Supplementation studies by Schuster et al. (1984) showed that at least 7.5 mg/day is required to prevent a decrease in maternal plasma pyridoxal phosphate (PLP) levels, while Apgar scores were higher for infants whose mothers had taken 7.5 mg or more supplemental pyridoxine than for infants of mothers who took 5 mg or less. The use of enzyme stimulation tests (EAST) or xanthurenic acid excretion after a tryptophan load as parameter, suggests that similar or even higher amounts are needed to maintain the results within the non-pregnant reference range (Coursin and Brown 1961; Lumeng et al. 1976). Remarkably, in our study with non-supplemented pregnant women (mean vitamin B6 intake 1.3 mg/day), plasma PLP levels at the end of pregnancy were in the same (low) range as, or even lower than, those reported by Roepke and Kirksey, without evidence of any adverse fetal development and with acceptable Apgar scores (>7 after 5 min) (van den Berg 1988a).

Table 2.3. Estimates of the vitamin B6 requirements in pregnancy

Criterion used	Estimate
Total amount transferred to the conceptus (10–20 mg)	+0.2 mg/d
Xanthurenic acid excretion after tryptophan loading (Coursin and Brown 1961)	15–20 mg/d
Plasma PLP, EGOT-stim test (Lumeng et al. 1976)	4–10 mg/d
Prevention of decrease in maternal plasma PLP (Schuster et al. 1984)	7.5 mg/d
Maximum cord plasma PLP (Schuster et al. 1984)	7.5 mg/d
Optimum Apgar score (at 1 min after birth) (Schuster et al. 1984)	7.5 mg/d
RDA (NAS/NRC 1980)	+0.6 mg/d (2.6 mg/d)

Assessment of the Nutritional Status in Pregnancy and Lactation

An important question in the assessment of the nutrient needs and the nutritional status of pregnant and lactating women is therefore whether the same parameters and interpretative criteria can be used as in studies with non-pregnant subjects. Nutritional status assessment can be approached at three different levels: (1) clinical examination, (2) dietary assessment, and (3) biochemical assessment. Clinical assessment does not seem very appropriate as for most micronutrients there are no specific clinical signs. Only for folate and iron has a causal relationship between a deficient status and (megaloblastic) anaemia been firmly established. For other micronutrients a relationship between nutrient deficiency and mortality and morbidity during pregnancy has been reported, but not confirmed (Hemminki and Starfield 1978). Dietary assessment is useful at the population level, but not for individuals. Moreover, for micronutrients the validity of this approach is questionable. Biochemical indices of the micronutrient status are, therefore, frequently used as more sensitive and specific indicators of malnutrition. Observational studies with pregnant women have indicated a relatively high incidence of abnormal values indicative of a marginal or deficient

B-vitamin or mineral status (for reviews see van den Berg and Bruinse 1983); Sauberlich 1978; Hambridge and Mauer 1978). Pregnancy-induced changes have been described for a variety of biochemical and physiological parameters. Many of these changes are in the direction of malnutrition. Thomson (1973) described the pregnant woman as follows: "She gains weight extremely rapidly. She may be sluggish and often complains of digestive upsets. Body temperature is slightly raised and pulse, respiration and basal metabolic rate are also increased. Haematological changes suggest anaemia. The erythrocyte sedimentation rate is characteristic of a chronic infection and the reduced level of serum albumin, often accompanied by oedema, suggests severe malnutrition". We may add to this that the hypovitaminosis and the higher incidence of abnormal nutritional status parameters are also suggestive of (multiple) micronutrient deficiency. Yet, at the end of all this, in nearly all cases, a perfectly healthy baby is born from a contented, healthy mother.

Many of the changes mentioned above are now considered as "physiological adjustments" of pregnancy, i.e., secondary changes in biochemical or physiological parameters occurring as part of the complex metabolic changes that enable the temporary symbiosis between mother and fetus, aiming at an optimum fetal development while maintaining maternal health.

It is impossible to summarize and discuss here all the pregnancy-induced changes in maternal physiology. Many excellent books and reviews are available on this subject, such as the standard work of Hytten and Leitch (1971) *The Physiology of Pregnancy*. I will only briefly mention some of the physiological adjustments which may be of relevance with respect to nutritional status assessment in pregnant and lactating women, and discuss some of the problems encountered in estimating the "nutrient cost" of pregnancy and lactation.

Physiological Adjustments in Pregnancy

Haematological Changes

Plasma volume starts increasing from the sixth week of pregnancy. Erythrocyte volume increases as well, but to a lesser extent, resulting in haemodilution. Haemodilution is often mentioned as an explanation of the hypovitaminosis of pregnancy as well as the lower levels of minerals and trace elements, such as serum iron and zinc. This may, indeed, hold for those micronutrients that are not bound to specific binding proteins and are not under metabolic (hormonal) control. Anttila and coworkers (1988) showed that serum zinc levels showed a similar pattern to serum albumin. However, haemodilution is definitely not a general causative factor (Bruinse et al. 1985; van den Berg 1988a). Binding capacity may even increase during pregnancy due to hormone-induced effects on concentration of binding proteins as demonstrated for the trancobalamins, folate-binding protein and ceruloplasmin (Da Costa 1974; Fernandez-Costa 1982).

Change in Organ Function; the Role of the Placenta

Gastrointestinal motility and tonus decrease during pregnancy, and so does secretion of hydrochloric acid and pepsin. These changes may affect nutrient absorption. However, for some nutrients, such as calcium, iron and zinc, increased absorption has been reported; vitamin absorption seems to remain unaffected.

Renal plasma flow and glomerular filtration rate increase by 30%–50%. However, this does not seem to result in a higher excretion of micronutrients. Micronutrient retention by the organism generally increases, but for folic acid an increased plasma clearance in pregnancy has been reported, resulting in an "extra" folate loss of about 15 μg/day (Fleming 1972).

Liver function remains relatively unaffected. In one study an increased folate turnover was reported due to a hormone-induced increase in microsomal enzyme activities (Davis et al. 1973). The placenta-derived increase in alkaline phosphatase activity may be one of the causative factors on the decrease in plasma PLP (van den Berg 1988a).

During pregnancy a "new" organ develops, the placenta. The placenta plays an important role in nutrient partitioning between mother and fetus (Munro 1983). For some B-vitamins as well as for some minerals and trace elements a relatively high retentive capacity of the placenta has been demonstrated (Munro 1983; Guigliani et al. 1985; Ramsay et al. 1983).

Change in Hormonal Balance; Metabolic Changes

During pregnancy massive changes in hormone production and hormonal balance occur. Increasing amounts of steroid and peptide hormones are produced by the developing fetoplacental unit. Many, if not all, of the physiological adjustments of pregnancy are mediated by the resulting change in hormonal balance. Evidence of direct hormonal effects on micronutrient metabolism is scarce. Studies with oral contraceptive (OCA) users indicate that concentrations of binding proteins, such as those for folate, vitamin B12, vitamin D and copper, increase (for a review, see Aftergood and Alfin-Slater 1980). Similar effects occur in pregnancy. Anabolic steroids generally increase vitamin (nutrient) retention and/or turnover (Beher and Gaebler 1951; Bamji et al. 1979). Serum levels of vitamin E and carotene are increased during pregnancy due to the increase in serum lipid content.

Hormonal changes result in a strong anabolic effect on maternal metabolism, the basal metabolic rate increases and the nitrogen balance becomes positive. Naismith (1980) showed evidence of the biphasic nature of protein metabolism: in the first half of pregnancy protein is deposited in maternal muscle tissue, while in the second half of pregnancy (the catabolic phase) this protein is re-used for the supply of amino acids to the fetus. Hytten and Leitch (1971) had already demonstrated the temporary deposit of fat to serve as an energy bank for the last part of pregnancy and for lactation. It is not yet clear what could be the consequences of these metabolic changes with respect to micronutrient partitioning. We have provided evidence of a parallel change in vitamin B6 stores in pregnant rats (van den Berg and Bogaards 1987). Also zinc retention (and requirement) are related to protein intake and nitrogen retention (Sandstead 1981). There can be no doubt that these physiological adjustments are of relevance for the selection and interpretation of biochemical parameters of nutritional status, and that changes in the parameters during pregnancy should be considered in the light of changes in maternal physiology. Many of the physiological adjustments of pregnancy are probably no longer, or of less relevance during lactation, although nutrient partitioning between milk output and maternal retention is under hormonal control (prolactin).

Establishing Interpretative Criteria for Pregnant and Lactating Women; Assessment of Risk Groups

In the longitudinal study performed some years ago in our Institute (van den Berg and Bruinse 1983; van den Berg 1988a) we demonstrated that for many vitamin and mineral status parameters a "spontaneous" recovery occurrs within six weeks after parturition, i.e., a normalization of vitamin blood levels to values within the "non-pregnant reference range" without dietary manipulation or supplementation. However, for serum folate, ferritin and plasma PLP, levels remained low. At 6 months postpartum the mean serum folate and ferritin level were still lower than those observed in early pregnancy. Serum zinc and selenium levels have been reported to show a similar pattern (Antilla et al. 1988). Based upon these observations we concluded that some of the pregnancy-induced changes in parameters of the vitamin and mineral status are indeed "secondary" and should be considered as "physiological adjustments of pregnancy". Also the more functional tests of the nutritional status, such as the enzyme stimulation tests, are affected, but to a lesser extent than are the blood levels.

One of the objectives of nutritional status assessment is the classification of individuals in terms of risk, and this requires a careful selection of interpretative criteria and cut-off points (Underwood 1986; van den Berg 1988b). Biochemical "normality", using cut-off points that are derived from the distribution of a certain parameter in a non-pregnancy reference population, does not take into account physiological effects or "adjustments" on parameters of the nutritional status. One can argue that the changes observed during a normal, uncomplicated pregnancy should be considered as normal physiology, and that the values measured can be used to define a reference range. Maintenance of maternal stores should be used as an additional criterion, i.e., a spontaneous recovery of the status parameters after pregnancy and/or lactation, to exclude depletion of maternal stores in the course of pregnancy. Maintenance of maternal stores can be inferred from the comparison of values of status parameters measured before pregnancy, or in its early stages, with those measured after pregnancy.

A better approach would be to relate the "acceptable" range to a specific physiological or metabolic response: which ranges of folate or iron levels are associated with an adequate haemoglobin level and haematological status; which range of blood nutrient level, or dependent enzyme activity is associated with an optimum mental and physical performance of mother and child, etc. (Sandstead 1981). Such an approach was chosen by Leibel (1982) who used the increase in haemoglobin content after iron supplementation as the parameter of the physiological response for validation of interpretation criteria of the iron status in children. However, for most micronutrients no suitable functional test seems as yet available and applicable in pregnancy.

Prevention of disease during pregnancy, "optimum" birth weight or infant development can also be used as criteria. However, for definite statements on micronutrient requirements in pregnancy and lactation, as well as on the criteria for classification into categories of risk, consensus should be reached on which criterion of adequacy is desired. For prevention of specific pregnancy-related disease, recommendations and dietary practice seem amply sufficient, at least in western societies, and this holds as well for birth weight and duration of pregnancy as the criterion, with a possible exception of folate requirements (Kristoffersen and Rolschau 1984). Using maintenance of (adequate) maternal stores as the criterion, our data (van den Berg 1988a) indicated that maintenance of adequate folate, iron and probably vitamin B6 stores may be difficult to achieve on a habitual diet, at any rate in the Netherlands. Women with short time intervals between successive pregnancies may, therefore, be at risk of developing an inadequate nutritional status for these nutrients.

Effect of Ageing on Micronutrient Status

Assessment of the Nutritional Status in the Elderly

One view of ageing is that it starts soon after conception, and includes all changes occurring between birth and death. This is evidently too broad a definition. "Elderly", "old" and "aged" are relative terms, without clear definitions. I will restrict myself to the effect of ageing on the nutritional status of the elderly (> 65 years). I will not discuss the role of nutritional status as a modifying factor of the ageing process.

There are numerous reports describing nutrient intake and nutritional status in the elderly (fore reviews, see Kirsch and Bidlack 1987; Kokkonen and Barrows 1986; Suter and Russell 1987). Generally, a lower nutrient intake, parallelling a lower energy intake, is observed when dietary data of elderly people are compared to those for younger age groups. This lower nutrient intake is generally associated with decreased serum and tissue nutrient levels. Supplementation with relatively modest amounts, in the same range as the RDAs, seems sufficient to "normalize" biochemical parameters (Kirsch and Bidlack 1987). Which nutrients are "at risk" seems dependent on various factors related to geographical differences, difference in age range and housing, i.e., institutionalized or living on their own (Kirsch and Bidlack 1987).

As an example, some recent data on nutrient intake and nutritional status from a nation-wide nutritional survey in the Netherlands among 500 apparently healthy elderly people between 65 and 79 years old, living on their own, are summarized in Table 2.4 (from Löwik et al. 1989a,b). Nutrient intake was generally lower as compared to mean values obtained with a younger age group. Comparison with recommended daily allowances indicate acceptable intake levels, except for vitamin B6. For men a mean vitamin B6 intake of 1.38 mg/day (17 μg/g protein) was estimated, and for women 1.15 mg/day (17 μg/g protein), while the RDA amounts are 1.5–2.0 mg/day (20 μg/g protein). Frank nutritional deficiencies were not observed. Mean values of biochemical parameters for elderly people were generally comparable with those of younger age groups, except that plasma PLP, serum folate and serum 25-hydroxy-vitamin-D contents were significantly lower. For most parameters a significant sex difference was observed, men generally having higher mean values than women.

Table 2.4. Dietary intake and nutritional status parameters for some selected micronutrients of elderly men (65–79 years) as compared to younger adult men (35 years) (from Löwik et al. 1989a, b)

Nutrient	Young (n=49)	Elderly (n=269)	Relative difference	Nutrient index[a]	Young (n=49)	Elderly (n=269)	Relative difference
Energy (MJ)	12.7	10.1	−29%				
Protein (g)	99	82	−19%				
Vit A (mg RE)	1.22	1.07	−13%	s-Ret	2.2	1.4	−36%
Vit D	–	–	–	s-25-OHD	71	40	−44%
Thiamin (mg)	1.37	1.11	−19%	ETK-AC[b]	1.10	1.10	0%
Riboflavin (mg)	2.18	1.70	−23%	EGR-AC	1.09	1.12	+ 3%
Vitamin B6 (mg)	1.60	1.38	−14%	p-PLP	48	27	−44%
Vitamin B12	–	–	–	s-B12	302	288	− 5%
Folate	–	–	–	s-Fol	10.0	6.6	−44%
Calcium (mg)	1.40	1.13	−20%	Ca-excr[c]	5.5	3.3	−40%
Iron (mg)	16.6	13.1	−22%	s-Fer	83	98	+19%

[a] Units nutrient indices: s-retinol (μmol/l); 25-OHD (nmol/l); p-PLP nmol/l; s-vitamin B12 (pmol/l); s-folate (nmol/l); Ca excretion (mmol/24h); s-Ferritine (μg/l).

[b] Data from Schrijver et al. (1985) (same laboratory as Löwik).

[c] Data from Binsbergen (1986) (same laboratory as Löwik).

What conclusions can be drawn from these data? Do they represent normal ageing phenomena or do they indicate an undesirable situation which requires action. Advertisements in newspapers and popular magazines suggest that vitamin and mineral supplements are essential for the elderly to maintain vitality and increase life expectancy. Indeed, the elderly are among the top consumers of supplements. Before discussing some of the aspects of physiological ageing that may be relevant with respect to nutritional status assessment in the elderly, I would like to make some remarks.

1. First of all, we should realize that ageing is associated with a significant increase in the incidence of diseases. Changes accompanying ageing are easily confused with those caused by the presence of secondary pathology. Most data concerning the process of ageing are derived from studies in which older age groups are compared with younger ones. However, many of these age-related changes should be re-evaluated as many of these age-related declines in function appear to have been due to undiagnosed diseases, rather than to ageing per se. This is one of the conclusions from the so-called SENIEUR collaborative study focusing on immunology and ageing (Ligthart et al. 1984). In the absence of disease, ageing per se does not necessarily lead to serious functional decline. However it is very difficult to discriminate between ageing and disease, because of the strong mutual interactions. Disease can cause malnutrition, but malnutrition can be a cause of disease. Atrophic gastritis, which may occur in up to 50% of the elderly, results in malabsorption of folic acid and vitamin B12 (Russell 1986). Achlorhydria may also be associated with impaired nutrient absorption, especially of protein-bound nutrients (Markkanen and Mustakallio 1968).

2. Some reports mention a relatively high alcohol consumption among the elderly (Barboriak and Rooney 1985). This may have the consequence that nutrient intake is even lower because a considerable part of the energy intake is derived from alcohol. In the presence of toxic ethanol levels, nutrient

metabolism may become impaired. In particular, some B vitamins (thiamin, vitamin B6, folate) and minerals (Zn, Mg) are at risk (for a review, see Yunice and Hsu 1986).

3. Elderly people as a group are the largest consumers of drugs, and in many cases (multiple) drugs are used for a prolonged time. Some drugs may seriously interfere with nutrient metabolism and need. Laxatives, diuretics and anticonvulsants should be of special concern because of their frequent and long-lasting use, and nutritional consequences, such as an impaired nutrient absorption or metabolism have been demonstrated for these drugs. Some drugs may affect taste acuity and appetite (for a review, see Roe 1985).

4. The elderly are a very heterogeneous group. As already mentioned, the presence of (undiagnosed) disease, but also a difference in age range may influence the results. Physiological features of persons between 60 and 70 years old can be very different from persons who are 80 or 90 (Schneider et al 1986). We will have to differentiate between independently living and institutionalized elderly. The first group is probably healthier, and mentally or physically less disabled. The latter group suffers more from chronic and intercurrent disease and medical intervention (drug use, surgery), which have been shown to be important determinants of malnutrition in elderly (DHSS 1979). Institutionalization in itself does not necessarily lead to impairment of the nutritional status (Sahyoun et al. 1988). Differences in socio-economic status may affect dietary intake and nutritional status, and poverty seems more common among the elderly than in other age groups.

The factors mentioned above may explain some of the observed differences in nutritional status and nutrient requirements of the elderly from those for younger people. However, for some nutrients "intrinsic" ageing effects on availability and/or metabolism have been demonstrated: vitamin A (Garry et al 1987), vitamin D (Armbrecht et al. 1984; Lips et al. 1987), vitamin B6 (Rose et al. 1976) and selenium (Ganapathy and Thimaya 1985).

Physiological Ageing

Age-induced changes in nutrient partitioning and metabolism are generally associated with changes in organ and physiological function. Interpretation of data from nutritional studies in elderly populations suffer, therefore, from similar difficulties as we came across in studies with pregnant women: i.e., do the changes observed reflect intrinsic ageing effects, can we use the same parameters and interpretative criteria as for younger adults, etc? Let us first consider some of the age-induced physiological changes that may be relevant with respect to nutrient requirement and metabolism. As for the physiological adjustments of pregnancy, I will only briefly discuss physiological ageing. For more extensive information the reader is referred to one of the many reviews on this subject (e.g. Watkin 1982; Chen 1986).

Body Composition; Metabolic Changes

Loss of lean body mass, reflected by a dramatic decrease in body water, is a basic characteristic of the ageing process in man. Total body water and intracellular water volumes decrease. Blood volume remains essentially unchanged. Fat and connective tissue content may increase with age. Bone mass and mineral content gradually decline, resulting in osteopenia. In women this process accelerates after the menopause. Data from the Baltimore Longitudinal Study (Shock et al. 1984) showed that energy intake diminishes steadily from 11.3 MJ at an age of 30 years to 8.8 MJ at 80 years. One-third of this decline can be accounted for by the reduced basal metabolism due to the decrease in muscle mass, the other two-thirds by a progressive reduction in physical activity (Shock et al. 1984). Protein synthesis and turnover show complex, age-related changes, albumin synthesis rate is reduced. The lower plasma albumin levels may result in reduced transport capacity for micronutrients. Although some reports have suggested that nitrogen requirements to maintain the nitrogen balance are higher for the elderly, there is as yet no sound evidence that, in the absence of catabolic states or malabsorption, nitrogen requirements are indeed higher for the elderly than for younger adults (for a review see McKay and Bond 1986). These changes in body composition and protein metabolism may affect vitamin and trace-element partitioning and turnover. Generally, vitamin B6, Zn and Ca requirements are related to protein requirements.

Organ Function

Extensive descriptions are available concerning age-related changes in organ and body functions, such as lung, liver and kidney functions and immune function. The most extensive studies on the effects of age on organ and tissue function in man are part of the Baltimore Longitudinal Study (Shock et al. 1984). In general, functional capacity decreases with age, i.e., changes are only apparent after challenge or stress. From a nutritional point of view the drop in kidney function, i.e., a decrease in renal blood flow and glomerular filtration rate is probably most relevant. In the kidney, activity of 1-alpha-hydroxylase, the enzyme responsible for formation of 1,25-dihydroxy-vitamin D, shows an age-dependent decrease (Armbrecht et al. 1984). The decrease in the ability of the skin to synthesize pre-vitamin D, due to the decrease in epidermal 7-dehydrocholesterol content, is of relevance as well (Holick 1986). In general, nutrient absorption capacity seems relatively unaffected by age, although for some nutrients, such as calcium and vitamin D, a decline in absorption with age has been described (DeLuca 1986). An increase in PLP hydrolysis, as a result from the increased serum alkaline phosphatase and/or liver hydrolase activity, has been mentioned as one of the factors explaining the lower plasma PLP levels in the elderly (see Suter and Russell 1987). As already mentioned, chronic and intercurrent disease, alcohol and drug use may all impair nutrient absorption.

The (Neuro-)endocrine System

Age-related changes in hormone functions are considered to be closely related to age effects on physiological function and the loss of metabolic control. The most significant changes are related to the decreased ability of endocrine systems, such as the adrenal-pituitary-hypothalamus axis, to respond to stress or stimulation (for a review, see Everitt and Walton 1988). The most apparent changes occur in the sex hormones and the glucocorticoid hormones. The loss of the menstrual cycle is considered as an important contributory factor in ageing osteopenia.

Nutrient Requirements for the elderly: Risk Groups

The ninth edition of the RDAs (NAS/NRC 1980) does not give specific recommendations for the elderly and has only proposed to reduce food energy intake. With respect to the micronutrients there seems at present no sound scientific basis for a recommendation to increase intake for healthy elderly people. However, Schneider et al. (1986) challenged the practice of establishing requirements based upon observed age-dependent changes in body composition and physiological function, considered as "normal". They suggested the formulation of dietary recommendations and interpretative criteria for (biochemical) parameters of the nutritional status, aiming at the maintenance of optimum physiological function and the prevention of age-dependent diseases. Rowe and Kahn (1987) introduced the term "successful ageing". In successful ageing extrinsic factors such as diet and lifestyle play a neutral or positive role on the ageing process, while in "usual ageing" these factors may exaggerate the basal or intrinsic ageing process.

Irrespective of whether nutrient requirements are different from those of younger age groups, the elderly have in general a higher risk of developing an inadequate nutritional status, and need to consume well-balanced diets of high nutrient density. There are several indications that an inadequate vitamin (nutrient) status may be a contributing factor in the development of impaired mental performance (Chomé et al. 1986) and immune response (Talbott et al. 1987). Exton Smith (1980) differentiated between two different categories of risk. One risk group is prone to develop malnutrition due to (primary) factors, such as low socio-economic status and physical inability interfering with food choice, capacity to purchase and consume foods, etc. A second category of risk includes the elderly with secondary causes of malnutrition such as impaired absorption, drug use and disease.

For adequate assessment of the nutritional status in elderly people we need more specific data on "intrinsic" age effects, i.e., age-related changes in nutrient metabolism and kinetics, not biassed by effects of pathology, drug use, etc. Mixed longitudinal studies such as the Baltimore Longitudinal Study are elegant, combining cross-sectional analysis with longitudinal data from the same subject, and so enabling distinctions to be made between ageing, cohort and secular effects.

Conclusions

General

Definite statements on micronutrient requirements, as well as on criteria for classification into categories of risk, can only be made after consensus has been reached on the desired "level of adequacy".

Pregnancy/Lactation

Using maintenance of maternal body stores during pregnancy as the criterion, the provision of iron, folate, and probably also vitamin B6 seems at risk for large groups of pregnant and lactating women.
Women with short time intervals between successive pregnancies are at risk of developing an inadequate nutritional status.

Ageing

There is, as yet, no sound basis for the conclusion that micronutrient requirements and metabolism for the elderly are different from those for younger adults, with the possible exception of vitamin D requirements.
Physical inability, a low socio-economic status, disease and drug use are among the main risk factors for malnutrition.

References

Aftergood L, Alfin-Slater RB (1980) Oral contraceptives and nutrient requirements. In: Human nutrition, vol 3B, Nutrition and the adult (micronutrients). Plenum Press, New York

Anttila P, Schmele S, Lehto J, Smell O (1988) Serum, zinc, copper, and selenium concentration in healthy mothers during pregnancy, puerperium, and lactation: a longitudinal study. In: Vitamins and minerals in pregnancy and lactation, Nestlé Nutrition Workshop series vol 16. Raven Press, New York

Armbrecht HJ, Prendergast JM, Coe RM (1984) Changes in calcium and vitamin D metabolism with age. In: Armbrecht HJ, Prendergast JM, Coe RM (eds). Nutritional intervention in the ageing process. Springer-Verlag, New York, pp 69–83

Bamji MS, Prema K, Lakshmi BAR et al. (1979) Oral contraceptive use and vitamin nutrition status in malnourished women: effects of continuous and intermittent vitamin supplements. J Ster Biochem 11:487–491

Barboriak JJ, Rooney CB (1985) Alcohol and its effects on the nutrition of the elderly. In: Watson RR (ed) Handbook of nutrition in the aged. CRC Press, Boca Raton, Florida, pp 215–248

Bauman DE, Currie BW (1980) Partitioning of nutrients during pregnancy and lactation: a review of mechanisms involving homeostasis and homeorhesis. J Dairy Sci 63:1514–1529

Beaton GH (1979) Nutritional needs of the pregnant and lactating mother. In: Hambraeus L, Sjölin S (eds) The mother/child dyad-nutritional aspects. Symposium of the Swedish Nutrition Foundation XIV. Almquist and Wiksell, Stockholm, pp 26–34

Beaton GH (1986) Towards harmonization of dietary, biochemical, and clinical assessments: the meanings of nutritional status and requirements. Nutr Rev 44:349–358

Beher WT, Gaebler OH (1951) Observations on niacin, riboflavin, allantoin, ascorbic acid and vitamin A during anabolism induced by hormones. J Nutr 41:447–457

Binsbergen JJ van (1986) Salt intake and blood pressure in a physicians practice (in Dutch). PhD Thesis, Maastricht, Rijksuniversiteit Limburg

Bruinse HW, van den Berg H, Haspels AA (1985) Maternal serum folacin levels during and after normal pregnancy. Eur J Obstet Gynec Reprod Biol 20:053–158

Chen LH (1986) Biomedical influences on nutrition of the elderly. In: Chen LH (ed) Nutritional aspects of ageing. CRC Press, Boca Raton, Florida, pp 53–67

Chomé J, Paul T, Pudel V et al. (1986) Effects of suboptimal vitamin status on behavior. Biblthca Nutr Dieta 38:94–103

Coursin DB, Brown VC (1961) Changes in vitamin B6 during pregnancy. Am J Obstet Gynecol 82:1307–1311

Da Costa M, Rothenberg SP (1974) Appearance of a folate binder in leucocytes and serum of women who are pregnant or taking oral contraceptives. J Lab Clin Med 83:207–214

Davis M, Simmons CJ, Dordoni B et al. (1973) Induction of hepatic enzymes during normal human pregnancy, J Obstet Gynaecol Brit Cwlth 82:374–81
DeLuca HF (1986) Significance of vitamin D in age-related bone disease. In: Hutchinson ML, Munro HM (eds) Nutrition and ageing. Academic Press, New York and London, pp 217–234
Department of Health and Social Security (DHSS) (1979) A nutritional survey of the elderly. Reports health soc subj no. 16. HMSO, London
Department of Health and Social Security (DHSS) (1977) The composition of mature human milk. Reports health soc subj no. 12. HMSO, London
Everitt AV, Walton JR (1988) Regulation of neuroendocrine aging. Interdisciplinary topics in gerontology, vol 24. Karger, Basel
Exton Smith AN (1980) Nutritional status: diagnosis and prevention of malnutrition In: Exton Smith AN, Caird FI (eds) Disorders in the elderly. John Wright, Bristol, p 66
Fernandez-Costa F, Metz J (1982) Levels of transcobalamins I, II and III during pregnancy and in cord blood. Am J Clin Nutr 35:87–94
Fleming AF (1972) Urinary excretion of folate in pregnancy. J Obstet Gynaecol Br Cwlth 79:916–928
Ganapathy SN, Thimaya S (1985) Selenium in the aged. In: Watson RR (ed) Handbook of nutrition in the aged. CRC Press, Boca Raton, Florida, pp 111–122
Garry PJ, Hunt WC, Bandrofchak JL (1987) Vitamin A intake and plasma retinol levels in healthy elderly men and women. Am J Clin Nutr 46:989–94
Giugliani ERJ, Jorge SM, Gonçalves AL (1985) Serum vitamin B12 levels in parturients, in the intervillous space of the placenta and in full-term newborns and their interrelationships with folate levels. Am J Clin Nutr 41:330–335
Hambridge KM, Mauer AM (1978) Trace elements. In: Laboratory indices of nutritional status in pregnancy. Committee on nutrition of the mother and preschool child. National Academy of Science, Washington DC
Hemminki E, Starfield B (1978) Routine administration of iron and vitamins during pregnancy: review of controlled clinical trials. Br J Obstet Gynaecol 85:404–410
Holick MF (1986) Vitamin D synthesis by the ageing skin. In: Hutchinson ML, Munro HM (eds) Nutrition and ageing. Academic Press, New York, pp45–58
Hytten FE, Leitch J (1971) The physiology of human pregnancy, Blackwell Scientific Publications, Oxford
Karra MV, Udipi SA, Kirksey A (1986) Changes in specific nutrients in breast milk during extended lactation. Am J Clin Nutr 43:495–503
Kennedy GC (1967) Ontogeny of mechanisms controlling food and water intake. In: Code CF (ed) Handbook of physiology, vol. 1, section 6. American Physiological Society, Washington, DC, p 337
Kirsch A, Bidlack WR (1987) Nutrition and the elderly: vitamin status and efficacy of supplementation. Nutrition 3:305–314
Kokkonen GC, Barrows CH (1986) Ageing and nutrition. Nutrition Int 2 (4): 205–212
Kristoffersen K, Rolschau J (1984) Vitamin supplements and intrauterine growth. In: Briggs MH (ed) Recent vitamin research. CRC Press, Boca Raton, Florida, pp 84–101
Leibel RL (1982) Studies regarding the impact of micronutrient status on behaviour in man; iron deficiency as a model. Am J Clin Nutr 35:1211–1221
Ligthart GJ, Corberand JX, Fournier C et al. (1984) Admission criteria for immunogerontological studies in man: the senieur protocol. Mech Ageing Dev 28:47–55
Lips P, Van Ginkel FC, Jongen MJM, Rubertus F, Van der Vijgh WJF, Netelenbox JC (1987) Determinants of vitamin D status in patients with hip fracture and in elderly control subjects. AM J Clin Nutr 46:1005–10
Lönnerdal B (1986) Effects of maternal dietary intake on human milk composition. J Nutr 116:499–513
Löwik MRH, Westenbrink S, Hulshof KFAM, Kistemaker C, Hermus RJJ (1989a) Nutrition and ageing: dietary intake of "apparently healthy" elderly. J Am Coll Nutr 8:347–356
Löwik MRH, Schrijver J, van den Berg H (1989b) Nutrition and ageing: Nutritional status of "apparently healthy" elderly. J Am Coll Nutr 9:18–27.
Lumeng L, Cleary RE, Wagner R et al. (1976) Adequacy of vitamin B6 supplementation during pregnancy: a prospective study. Am J Clin Nutr 29:1376–1383
Markkanen T, Mustakallio E (1963) Absorption and excretion of biotin after feeding minced liver in achlorhydria and after partial gastrectomy. Scand J Clin Lab Invest 15:57–61
McKay MJ, Bond J (1986) Protein and amino acids. In: Chen LH (ed) Nutritional aspects of ageing. CRC Press, Boca Raton, Florida, pp 173–194
Munro HN (1983) The placenta in nutrition. Annu Rev Nutr 3:97–124

Naismith DJ (1980) Endocrine factors in the control of nutrient utilisation in pregnancy. In: Aebi H, Whitehead R (eds) Maternal nutrition during pregnancy and lactation. Nestlé Found publ series no. 1, Hans Huber, Bern, pp 16–26
National Academy of Sciences – National Research Council (1980) Committee on Dietary Allowances of the Food and Nutrition Board, 9th revised ed. National Academy of Sciences, Washing DC
Ramsay VP, Neumann C, Clark V et al. (1983) Vitamin cofactor saturation indices for riboflavin, thiamin, and pyridoxine in placental tissue of Kenyan women. Am J Clin Nutr 37:969–973
Roe DA (1985) Drug effects on nutrient absorption, transport, and metabolism. Drug Nutr Interactions 4:117–136
Roepke JLB, Kirksey A (1979) Vitamin B6 nutriture during pregnancy and lactation. I. Vitamin B6 intake, levels of the vitamin in biological fluids and condition of the infant at birth. Am J Clin Nutr 32:2249–56
Rose CS, Gyorgy P, Butler M et al. (1976) Age differences in vitamin B6 status of 617 men. Am J Clin Nutr 29:847–53
Rowe JW, Kahn RL (1987) Human ageing: usual and successful. Science 237:143–149
Russell RM (1986) Implications of gastric atrophy for vitamin and mineral nutriture. In: Hutchinson ML, Munro HM (eds) Nutrition and ageing. Academic Press, New York and London, pp 59–70
Sahyoun NR, Otradovec CL, Hartz SC et al. (1988) Dietary intakes and biochemical indicators of nutritional status in an elderly, institutionalized population. Am J Clin Nutr 47:524–533
Sandstead HH (1981) Methods for determining nutrient requirements in pregnancy. Am J Clin Nutr 34:697–704
Sauberlich HE (1978) Vitamin indices. In: Laboratory indices of nutritional status in pregnancy. Committee on nutrition of the mother and preschool child. National Academy of Science, Washington DC
Schneider EL, Vining EM, Hadley ED, Farnham SA (1986) Recommended dietary allowances and the health of the elderly. The New Engl J Med 314:157–160
Schrijver J, Van Veelen BWC, Schreurs WHP (1985) Biochemical evaluation of the vitamin and iron status of an apparently healthy dutch free-living elderly population. J Vitam Nutr Res 55:337–349
Schuster K, Bailey LB, Mahan CS (1984) Effect of maternal pyridoxine HCL supplementation on the vitamin B6 status of mother and infant and on pregnancy outcome. J Nutr 114:977–988
Shock NW, Greulich RC, Andres R et al. (1984) Normal human ageing: the Baltimore Longitudinal Study od ageing. NIH Publ. No 84–2450, NIH, Washington DC
Smith JC, Hsu JM (1982) Trace elements in ageing research: Emphasis on zinc, copper, chromium, and selenium. Nutritional approaches to ageing research. CRC Press, Inc., Boca Raton, Florida, pp 120–134
Suter PM, Russell RM (1987) Vitamin requirements of the elderly. Am J Clin Nutr 45:501–12
Talbott MC, Miller LT, Kerkvliet NI (1987) Pyridoxine supplementation: effect on lymphocyte responses in elderly persons. Am J Clin Nutr 46:659–664
Thomson AM (1973) In: Hytten FE, Lind T (eds) Diagnostic indices in pregnancy. CIBA-Geigy, Basel, p 1
Underwood BA (1986) Evaluating the nutritional status of individuals: a critique of approaches. Nutr Rev Suppl [May 1986] 213–224
van den Berg H, Bruinse HW (1983) On the role of nutrition in normal human pregnancy. PhD Thesis, University of Utrecht, The Netherlands
van den Berg H, Bruinse HW (1984) Assessment of the nutritional status of women during normal pregnancy. In: Fidanza F (ed) Proc. GEN-Workshop, Perugia, Italy
van den Berg H, Bogaards JJP (1987) Vitamin B6 metabolism in the pregnant rat: effect of progesterone on the (re)distribution in maternal vitamin B6 stores. J Nutr 117:1866–1874
van den Berg H, Bruinse HW (1988) Vitamin requirements in normal human pregnancy. Wld Rev Nutr Diet 57:95–125
van den Berg H (1988a) Vitamin and mineral status in healthy pregnant women. In: Berger H (ed) Vitamins and minerals in pregnancy and lactation. Nestlé nutrition workshop series vol 16, pp 93–108. Nestec Raven Press, Vevey, New York
van den Berg H (1988b) Assessment of the B vitamin status: some biochemical and clinical aspects. In: Souverijn JHM, Den Boer NC (eds) Clinical chemistry: an overview. Plenum Press, New York, pp 545–556.
Van Zoeren-Grobben D, Schrijver J, van den Berg H, Berger HM (1987) Human milk vitamin content after pasteurisation, storage, or tube feeding Arch Dis Child 62:161–165
Watkin DM (1982) The physiology of ageing. J Clin Nutr 36:750–758
Yunice AA, Hsu JM (1986) Alcohol. In: Chen LH (ed) Nutritional aspects of ageing, vol 2. CRC Press, Boca Raton, Florida, pp 19–71

Chapter 3

Critical Appraisal of Current Approaches Towards Micronutrient Requirement

G. Brubacher

Introduction

All recommendations for dietary intake (RDI), allowances (RDA) or amounts (RDA) are based on scientific data and reasonable judgements. Whereas scientific data are universally valid and may change only if new scientific evidence becomes available, reasonable judgements depend on the opinion of single scientists or scientific committees, and may vary according to the economic or nutritional conditions or the mode of thinking from country to country. This fact is the main reason that RDIs or RDAs differ between various countries. The scientific data on which RDIs or RDAs are based are often not known exactly and have to be estimated. These estimates may vary according to the methods used and may also be a reason for the variability of RDIs and RDAs.

The Analytical Logical Approach

This approach has been used by several scientists and scientific bodies, among them WHO/FAO (1989) and has been critically discussed by Brubacher (1989).

The basic scientific facts and reasonable judgements, which are inherent in this system, are the following:

1. We cannot speak about requirement, pure and simple, but only about requirement for a certain function. For example we distinguish between minimal requirement for maximal growth or for maximal reproduction, or for avoidance of deficiency symptoms.

In general the exact figures of this minimal requirement are not known and have to be estimated.

2. A reasonable judgement in affluent societies is to choose the minimal

requirement of a micronutrient to avoid the first sign of deficiency as a base for the RDA.

This first sign of deficiency may be a functional impairment, such as depressed working capacity or mental properties, or biochemical changes, such as elevation of blood pyruvic acid level or prolongation of blood clotting time, or finally morphological changes, like hypersegmentation of granulocytes or disappearance of goblet cells. Let us, therefore, call this requirement functional requirement.

For most micronutrients these very first deficiency symptoms are not yet known and, if known, the exact figure of the minimal requirement for avoidance of the symptoms is not known with the desirable accuracy. Thus, in many cases, estimates have to replace exact figures. There is, therefore, a great need for further research.

In less favourable situations, as may be the case in developing countries, the first symptom of clinical deficiency, and not the minimal requirement for avoidance of the very first sign of deficiency may be adopted. This minimal requirement was called by FAO/WHO (1989) basal requirement. This procedure is not unusual. For instance, the minimal cost of living differs from country to country according to the economic power.

3. In general, recommendations are directed towards population groups and not towards single persons. Therefore the basis for recommendations should be set by the distribution of requirements in population groups. This distribution is not known in most cases. It should be stressed that the distribution is probably in most cases not Gaussian but inclined to the left, as it is in the case of iron. Despite this fact most estimates are based on the assumption of a normal distribution. Here again, further research is needed.

4. A reasonable judgement consists in choosing the 95th percentile of the requirement distribution as a basis for recommendations. In developing countries, again, a lower percentile may be adequate. In cases where the distribution is very skew, problems may arise since an appreciable part of the population may not be able to cover their requirement for micronutrients from food and may require enriched food or dietary supplements. In making recommendations a lower percentile may be accepted and, at the same time, those exceeding this percentile may be recommended to cover their requirements by enriched food or supplements.

5. Since, for one reason or another the supply of micronutrients from food may decrease or cease, e.g., by slimming or loss of appetite caused by disease, micronutrient intake should be adequate to build reserves. Reasonable or normative judgements should define the time-span in question. In tropical countries during the rainy season, of which the duration is about three months, the micronutrient supply is minimal and it is reasonable to choose a time-span of three months in these regions. It is not unreasonable to accept the same time-span in affluent societies.

6. The magnitude of these reserves may be estimated from kinetic data, which also allow us to calculate the daily intake needed to maintain them. This intake may be called the normative storage requirement. In most cases kinetic data are not yet available to make accurate estimates and it is again a challenge for further research.

7. For the normative storage requirement, as in the case of the functional requirement, the distribution in a population group is of major importance.

Where this distribution is not known or cannot be estimated further research is needed.

8. A reasonable judgement in affluent societies may be to accept the normative storage requirement for the 95th percentile as the recommendation for the daily intake of a micronutrient.

In summary the following questions have to be answered for constructing the figures of RDAs:

1. What are the first subclinical signs of micronutrient deficiency for specified micronutrients?
2. What is the minimal intake to avoid this first sign and how big is the corresponding body pool?
3. How is this minimal requirement and the corresponding body pool distributed in a certain population group, and what value has the 95th percentile?
4. What intake corresponds to a body store which allows an interruption for three months of the micronutrient supply to the body without the appearance of the first sign of subclinical vitamin deficiency?
5. How is this intake and the corresponding body pool distributed and what value has the 95th percentile?

Despite the fact that most of these questions can only be answered by estimates, figures derived by this system may give a fair estimate of the RDA for a series of micronutrients. However the main value of the system consists in precise formulation of the problems which need further research, and in the exact discrimination between estimates which are based on scientific data and reasonable judgement, which is often based on opinions of single scientists or scientific committees.

The Epidemiological Approach

A second approach to construct RDAs uses the epidemiological method. In the simplest case a population group is looked for in which frank deficiency symptoms occur. In this group the intake of the micronutrient and the occurrence of the deficiency symptom for the single subjects is investigated and by the regression calculation the cut-off point for the micronutrient intake is determined above which no symptom is any more observed. Since the regression line is non-linear, reasonable estimates for the form of this line have to be made. In the past, population groups of different countries have often been chosen and the mean micronutrient related to the prevalence of the deficiency symptom in these groups.

As long as the causal relationship between the micronutrient intake is given, the basis of the epidemiological method is safe. But in many cases there is only a statistical relationship between the micronutrient intake and certain health parameters, without the knowledge whether this relationship is due to a direct causal relationship or to a third factor, which is related to the health parameter. It depends on the opinion of the single scientist or the scientific committee involved in the construction of RDAs whether or not such relationships are admitted. A proof of a causal relationship can only be given by intervention trials. But this type of trial is very costly and of long duration.

Conservative scientists are inclined to neglect such relationships. Despite this attitude, from a public health point of view it seems prudent to take such a relationship into consideration.

To give an example: there exists a well documented relationship between vitamin A deficiency symptoms and intake of vitamin A by children of third world countries. Sommer, Tarwotijo and West (1988) pointed out that children with vitamin A deficiency symptoms had a higher mortality than similar children without these symptoms, and they were able to demonstrate in an intervention trial that the relationship between excess mortality and vitamin A intake is causal. Therefore this relationship should be taken into consideration by construction of RDAs for vitamin A.

Other relationships of this kind are those between intake of vitamin C and plasma cholesterol (Ginter and Robek 1981), between vitamin E intake and ischaemic heart disease (Gey, Brubacher and Stähelin 1987) and between calcium intake and osteoporosis (Albanese 1983).

In epidemiological studies the micronutrient intake is often not directly measured but derived from measurement of plasma levels or other biochemical parameters. An example for this kind of procedure is given by Suboticanec-Buzina et al. (1984), where the relationship between physical working capacity and vitamin C intake is discussed.

The Balance Approach

In a third approach results of balance studies are used. In a simple model the cut-off point is looked for where micronutrient intake equals the sum of micronutrient excretion and degradation. With smaller intake the balance becomes negative and with higher intake the balance becomes positive. With minerals the problem becomes even simpler, since there is no degradation. The cut-off point is equal to the requirement and by considering the requirement distribution in a population group recommendations can be derived similar to the procedure described below. Nordin et al. (1987) have described how exact figures for calcium requirement can be found, by this procedure.

However it seems that the model used is too simple to reflect the biological process. It is true that for low micronutrient intake the balance becomes negative and that this intake is certainly below the requirement. By increasing the intake, a point of equal balance is reached. This point corresponds to a certain magnitude of the body pool and often to a certain plasma level of the micronutrient in question. However with further increase of the intake the balance remains equal corresponding to a higher body pool or a higher plasma level. Only when the intake is increased dramatically can excretion and degradation no longer keep pace with intake and undesirable effects arise. The relationships can be vizualized by hydrodynamic models. It is clear that degradation and excretion is minimal at the lowest level where balance is reached, but it cannot be concluded that the intake at this level is necessarily the requirement. By using this method it is important to fix a given body pool or a given plasma level by reasonable judgement or with the help of scientific data not coming from balance studies or kinetic properties. Kallner (1981) has given an example of an even more complicated model by calculation of the vitamin C requirement assuming a body pool of 1500 mg ascorbic acid or a plasma level between 0.8 and 0.9 mg/100 ml.

The Pragmatic Approach

When this method is used for calculation of RDAs, the micronutrient intake is studied in a large, apparently healthy population group. It is assumed that the intake in such a group is safe and adequate. The committee involved in the publication of the 1980 edition of the American RDAs (National Research Council 1980) decided not to calculate figures for RDAs from these figures of safe and adequate intakes but to publish them as such. The main point of criticism is that it is not known whether an apparently healthy population is really healthy or whether the naturally occurring morbidity or mortality in such a population is caused, to a certain degree, by marginal micronutrient deficiencies. Therefore in constructing RDAs it would be prudent to rely on the higher figures.

Special Considerations

It remains to consider several special cases; for instance, the need to enhance the recommendation for the amount of the daily vitamin C intake in regions with high prevalence of iron deficiency, the need to enhance the recommendations for the amount of the daily vitamin E intake in situations where the intake of polyunsaturated fatty acid is high, the property of ascorbic acid to block the formation of nitrosamines in the stomach. Thus special circumstances have to be taken into account for some nutrients which cannot be discussed in this paper, but which are nevertheless important.

Bioavailability

In general, RDAs do not take into consideration the differences in bioavailability of naturally occurring micronutrients. This problem is not yet solved. Conceivably bioavailability could be included in food composition tables. But in general it is not the bioavailability of a micronutrient in an isolated food item but the bioavailability in the ready-to-eat dish which is of interest and which might be very different. Inclusion of bioavailability in food tables would therefore be misleading. It is also conceivable to elaborate factors with which the RDAs should be multiplied, but this approach also presents many difficulties.

The Concept of Relative Nutritional Density

RDAs may be used for food labelling purposes or more generally for characterization of the nutritional properties of a certain food item. In such cases RDAs have to be related to some food properties.

We can calculate what fraction of the RDA is contained in 100 g of the food item in question, in a serving or in a daily portion. Such figures may sometimes be misleading. Let us take a food item which would supply 5% of a certain RDA in 100 g and let us suppose that this food item contains 200 kcal/100 g. 2000 kcal of such a food item supply only 50% of RDA; so this food must be considered a poor micronutrient source. On the other hand, if the food item containing 5% of this RDA per 100 g supplied only 20 kcal/100 g, 400 kcal would be necessary to supply 100% of the RDA. Such a food item could be classified as a rich source of the micronutrient in question. So logically the micronutrient content should be related to the energy content of a food item (nutritional density).

It is generally assumed that, above a certain cut-off point (at least for vitamins), the requirement increases proportionally to the energy intake. Let us assume that this cut-off point lies at 10 000 kJ, an amount which is needed by a moderately active young male adult to maintain his body weight. Let us further calculate the micronutrient content of a food item per 10 000 kJ, which ratio corresponds to the nutritional density of the food item, and let us divide this figure by the RDA in question: we obtain a figure which could be called relative nutritional density. We can conclude that everybody who eats 10 000 kJ or more a day with a mean relative nutritional density of 1, covers his RDA. If such a person covers half of his daily ration with refined food devoid of micronutrients the rest of his daily ration should have a relative nutritional density of 2. If he has for a certain micronutrient only a few food items, which deliver only a quarter of his energy need, these food items should have a relative nutritional density of 4.

Modern lifestyle often results in a reduction in energy needs; in slimming diets, too, the energy intake is less than 10 000 kJ. Finally, elderly people eat less than 10 000 kJ each day. It is generally assumed that in these situations the micronutrient requirement is not reduced and thus the RDA should not be changed. This means that in these cases the average relative nutritional density should be greater than 1. For instance, with a daily intake of 8000 kJ the average relative nutritional density should be 1.2 or with a daily intake of 5000 kJ the average relative nutritional density should be 2.

The concept of relative nutritional density is not useful for food supplements, because the figures would tend toward infinity, since the energy content tends to zero. In such supplements the micronutrient content should be declared by fractions of the RDAs.

Summary

Recommendations for the daily allowances (RDA) or the amount of daily intakes (RDI) of micronutrients are based on scientific data and reasonable judgements. Whereas scientific data may only change if new scientific evidence becomes available, reasonable judgements may differ according to the political or economic situation of a country or according to the opinion of a single scientist or a scientific committee involved in elaboration of RDAs or RDIs. This difference in judgements is the main reason of the variability of RDAs between the various countries.

In principle there are four different approaches for construction of RDAs or RDIs, which may be called the analytical logical, the epidemiological, the balance, and the pragmatic approach. Beside these four methodological approaches there are some special physiological properties of most micronutrients which have to be taken in consideration.

By translating the figures of the RDAs or RDIs to practical problems it seems that the concept of the relative nutritional density is of value. The relative nutritional density corresponds to the micronutrient content per 10 000 kJ of a food item divided by the RDA or RDI of the micronutrient in question. The figure obtained in this manner allows a judgement to be made whether a food item is a poor or a rich source of a certain micronutrient and whether such a food item is suitable to be introduced in a low caloric diet.

References

Albanese AA (1983) Calcium nutrition throughout the life cycle. Bibltheca Nutr Dieta 33: 80–99

Brubacher GB (1989) Scientific basis for the estimation of the daily requirements. In: Walter P, Stähelin H, Brubacher G (eds) Int J Vitam Nutr Res [Suppl] 30:3–11

Gey KF, Brubacher GB, Stähelin HB (1987) Plasma levels of antioxidant vitamins in relation to ischemic heart disease and cancer. Am J Clin Nutr 45:1368–1377

Ginter L, Robek P (1981) The influence of vitamin C on lipid metabolism. In: Counsell JN, Hornig D (eds) Vitamin C, ascorbic acid. Applied Science, London New Jersey, pp 299–347

Kallner A (1981) Vitamin C, man's requirement. In: Counsell JN, Hornig D (eds) Vitamin C, ascorbic acid. Applied Science, London New Jersey, pp 63–73

National Research Council (1980) Recommended dietary allowances. National Academy of Science, Washington DC

Nordin BEC, Polley KJ, Need AG, Morris HA, Marshall D (1987) The problem of calcium requirement. Am J Clin Nutr 45:1295–1304

Sommer A, Tarwotijo K, West KP (1988) Impact of vitamin A deficiency on infant and childhood mortality. In: Berger H (ed) Vitamins and minerals in pregnancy and lactation. Nestlé nutrition workshop series vol 16. Raven Press, New York, pp 413–419

Suboticanec-Buzina K, Buzina R, Brubacher G, Sapunar J, Christeller S (1984) Vitamin C status and physical working capacity in adolescents. Int J Vitam Nutr Res 54:55–60

WHO/FAO (1989) Report of joint FAO/WHO expert group, requirements of vitamin A, iron, folates and vitamin B12. WHO, Geneva

Chapter 4

Identification of Those Micronutrients Most Likely To Be Insufficient as the Result of Habitual Low Energy Intake

J. P. Mareschi

The technical, scientific and economic progress of the last decades in the industrialized countries has led to changes in lifestyle including, of course, changes in food habits (Haeusler 1985, 1986).

Among the causes of an inadequate intake of minerals and vitamins are a reduction of the intake of foods containing the nutrients and a reduction of the energy intake. It is generally admitted that the most frequent dietary errors observed in different industrialized countries are caused by the factors listed in Table 4.1. These trends have been confirmed in France (Dupin et al. 1984; Anonyme 1986), in Italy (Fidanza et al. 1984) and in England (Nelson 1986). In addition there is an increase in the consumption of low-energy foods to avoid overweight (Birch and Lindley 1988).

Unfortunately in many countries, including France, few surveys have been carried out to relate food intake and the relative nutritional status; in France most available data are about iron. We tried to find a method which could give us an estimate of the intake despite the lack of large, detailed investigations. The balanced diet is the goal. We determined the mineral and vitamin intakes supplied by a balanced diet related to the energy intake level. The content of these diets was compared with results from available surveys. The aim of this paper is to present and discuss the validity of the method, based on the content of minerals and vitamins in the balanced diet, as an indicator of the adequacy of the intake of these nutrients by the population of the industrialized countries.

Materials and Methods

Different menus, balanced for protein, lipids and carbohydrates, were designed by Professor Debry's team from the Human Nutrition Centre at Nancy University

Table 4.1. Frequent dietary errors in industrialized countries

Food intake			
Excess	Deficit	Intake of other components	Life habits
Lipids	Cereals[a]	Cigarette smoking	Stress
Alcohol	Legumes	Oral contraceptive	Sedentary
Protein	Fruits[b]	Drugs (laxatives, antidepressants,	
(from meats, cheese)	Vegetables[b]	tranquillizers etc)	
Common salt	Fish		
Coffee	Milk		
Sugars	Water		

[a] Often too refined and deficient in fibre.

[b] Often too low to supply enough vitamins, minerals and dietary fibre.

(Debry 1986). The energy distribution was 14.1% for proteins, 32% for lipids and 53.9% for carbohydrates.

These balanced menus were established for the following energy levels: 1500 (3.5 MJ), 2000 (4.78 MJ), 2500 (6 MJ), 3000 (7 MJ) and 3500 (8.4 MJ) kcal/day; they do not take into consideration beverage consumption.

Amongst this range of energy levels, we have been particularly interested in those which are close to the energy intake of the average French adult (Dupin 1981), for men (2700 kcal/d), women (2000 kcal/d) and lactating or pregnant women (2500 kcal/d).

The vitamins and minerals supplied by the three menus in the study were established according to the composition table of Souci et al. (1986) or, when the information was not available there, according to that of Paul et al. (1978). The calculations were made on raw foods for minerals and on cooked foods for vitamins.

We did not consider the data for selenium and chromium because the values taken from the tables differed too much. The elements ingested at levels greater than need (sodium and potassium) were not analyzed. The true intake of pantothenic acid is unknown because of biosynthesis in the gut and the analytical method is unreliable. Niacin intake was underestimated because the niacin-equivalent coming from tryptophan ingestion was not included. Biotin was not studied because many data are missing in the food composition tables and the contribution from biosynthesis in the gut is not known.

Once calculated, mineral and vitamin intakes were compared with the recommended intakes for the three types of the considered population.

The retinol and provitamin A components of vitamin A intake (retinol equivalents) were expressed separately. We thought it wise to consider that the retinol intake has to be at least 50% of the recommended vitamin A intake.

Anderson et al. (1982) have established a relation between the level of mineral intake as compared with the recommended allowances and the probability that the population has an adequate or insufficient supply of minerals. The level of 80% of the recommended allowances appeared as a critical level below which there is the probability of deficiency. For this level of 80%, Anderson showed that there is a 30% probability that the population will receive amounts of minerals below the recommended allowances.

Results

The mineral and vitamin intakes supplied by the balanced diet are expressed as percentage of the recommended intake for men, women, and pregnant and lactating women (Table 4.2 and Figs. 4.1 and 4.2).

Table 4.2. Vitamins, minerals and trace elements supplied by a balanced diet, and the percentages of recommended daily amounts supplied by balanced diets of various calorie contents

Minerals	a	a	a	b	a	a	a	b	b
Minerals and trace elements	Ca	P	Mg	Cu	Fe	Zn	I	F	Mn
RDA for women (2000 kcal)	800	800	350	2.5	18	15	0.12	2.7	3.8
RDA for pregnant and lactating women (2500 kcal)	1100	1200	400	2.5	20	22.5	0.14	2.7	3.8
RDA for men (2700 kcal)	800	800	350	2.5	10	15	0.12	2.7	3.8
Balanced diet (2000 kcal)	893	1247	302	1.8	13.7	11	0.1	0.6	3.2
Balanced diet (2500 kcal)	1076	1482	348	2.3	15.6	12.8	0.12	0.8	4.1
Balanced diet (2700 kcal)	1162	1600	376	2.4	16.8	13.8	0.13	0.8	4.5
Percentage of RDA									
Balanced diet 2000/RDA	112	156	86	72	76	73	83	22	84
Balanced diet 2500/RDA	98	124	87	92	78	57	86	30	108
Balanced diet 2700/RDA	145	200	107	96	168	92	108	30	118

Vitamins	ad	ae	a	a	a	a	a	a	a	a
Vitamins	Vit. A (μg)	Toco. (mg)	Vit.B1 (mg)	Vit.B2 (mg)	Vit.B6 (mg)	Vit.B12 (mg)	Vit.C (mg)	Vit.PP (mg)	Panto-thenic acid (mg)	Folic acid (mg)
RDA for women (2000 kcal)	800	10	1.3	1.5	2	0.003	80	15	8.5	0.4
RDA for pregnant and lactating women (2500 kcal)	1300	10	1.8	1.8	2.5	0.004	90	20	8.5	0.65
RDA for men (2700 kcal)	1000	10	1.5	1.8	2.2	0.003	80	18	8.5	0.4
Balanced diet (2000 kcal)	2181	6.42	0.97	1.28	1.53	0.004	95	11.3	5	0.34
Balanced diet (2500 kcal)	2241	6.78	1.15	1.45	1.79	0.005	97	13.2	6.1	0.38
Balanced diet (2700 kcal)	2425	7.32	1.24	1.57	1.94	0.005	104	14.3	6.6	0.41
Percentage of RDA										
Balanced diet 2000/RDA	273	64	75	85	77	133	119	75	59	85
Balanced diet 2500/RDA	172	67	64	81	72	125	108	66	72	58
Balanced diet 2700/RDA	243	73	83	87	88	167	130	79	78	103

[a] Dupin (1981).
[b] RDA (1980) National Academy of Sciences, Washington DC.
[c] Calculated by balanced diet of 2500 kcal.
[d] Provitamin A represents 87% of total vitamin A in balanced diet of 2000 kcal, 85% of total vitamin A in balanced diet of 2500 kcal, 84% of total vitamin A in balanced diet of 2700 kcal.
[e] Tocopherol is calculated as alpha tocopherol equivalent.

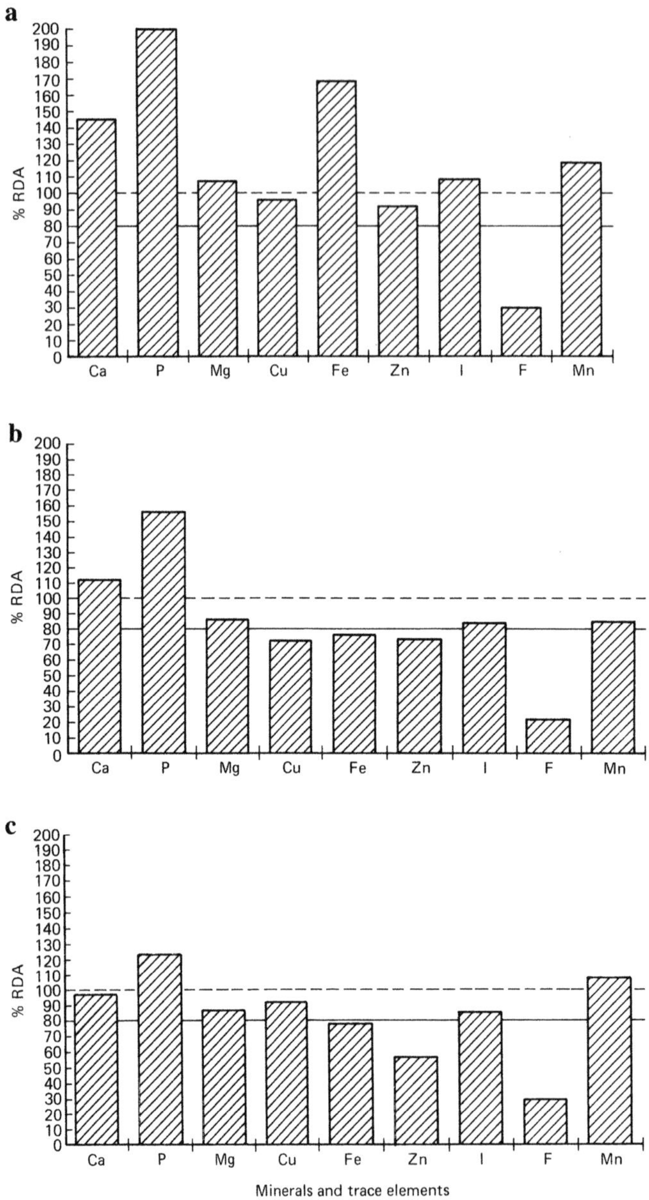

Fig. 4.1a,b,c. Percentages of recommended daily intakes of minerals and trace elements in balanced diets. **a** In men (2700 kcal); **b** in women (2000 kcal); **c** in pregnant women (2500 kcal).

In decreasing order, the elements with the risk of a deficient intake (intakes lower than 80% of that recommended) are shown in Table 4.3. The results show that, even with a balanced diet in which "empty calories" intake is reduced to a minimum and a diversified diet is supplied, it becomes difficult to reach the recommended intake below 2700 kcal/day.

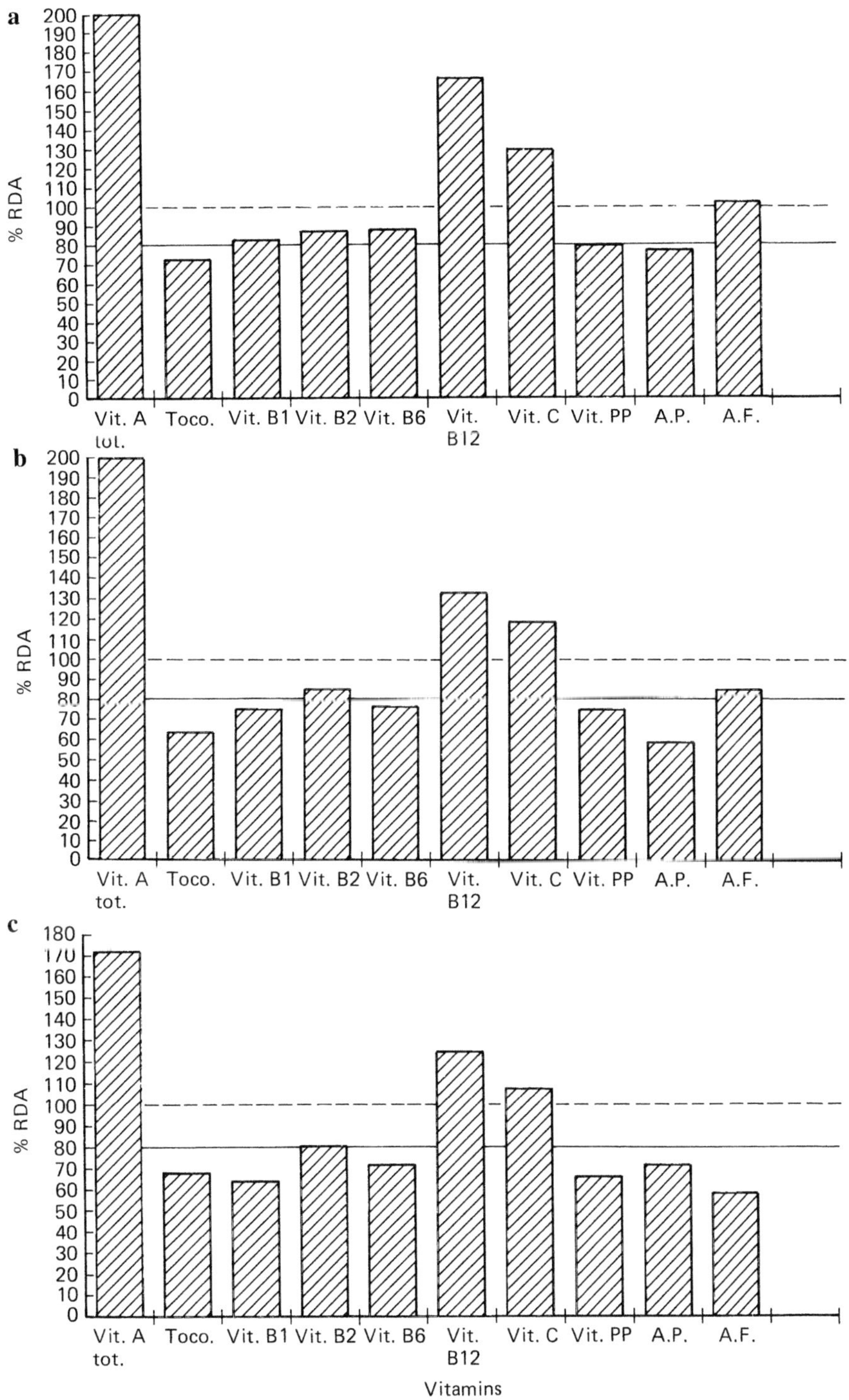

Fig. 4.2a,b,c. Percentages of recommended daily intakes of vitamins in balanced diets. **a** In men (2700 kcal); **b** in women (2000 kcal); **c** in pregnant women (2500 kcal).

Table 4.3. Micronutrients at risk of being deficient in balanced diets (Mareschi et al. 1987)

	Minerals	Vitamins
Adult men (2700 kcal)	Fluorine	Retinol free tocopherol>pantothenic acid
Adult women (2000 kcal)	Fluorine>iron>copper>zinc> manganese	Retinol free pantothenic acid> tocopherol> vitamin B1> vitamin B6
Pregnant, lactating women (2500 kcal)	Fluorine>zinc>iron	Retinol free folic acid>vitamin B1> vitamin B6>tocopherol>

Discussion

Criticism of the Use of a Balanced diet for the Detection of Minerals and Vitamins at Risk of being Deficient for the Adult Population

Firstly, only a small part of the population consumes this type of diet. Secondly, intakes are overestimated (quality and quantity) relative to the actual consumption of the population. Finally, the beverages are not taken into consideration.

Validity of the Use of a Balanced Diet for the Detection of Minerals and Vitamins at Risk of being Deficient for the Adult Population

The balanced diet supplies slightly more of the vitamins and minerals compared with the "observed" diet (Mareschi et al. 1987b, Aubree et al. 1988). The elements at risk in the common diet still have the same risk with the balanced diet (Figs. 4.3, 4.4).

The minerals and vitamin intake profiles are schematically comparable if we consider the "balanced" diet and the "observed" one studied by the "purchasing method". Also, food investigations and surveys on nutritional status in different industrialized countries (in France only iron has been subject to such studies) give similar results to those supplied by the analysis of the balanced diet (Couzy et al. 1988, Aubree et al. 1988).

Conclusion

Since the mineral and vitamin intakes provided by a balanced diet are still inadequate and the literature does not give enough data about the intake of these nutrients, we think that the use of the balanced diet is quite a good method. One of the reasons that the minerals and vitamins are inadequate is related to the energy intake, which is more important than whether the diet is balanced or not. The analysis of the balanced diet seems to be a good indicator of minerals and vitamins with a high risk of deficiency in the common diet. Mareschi et al.

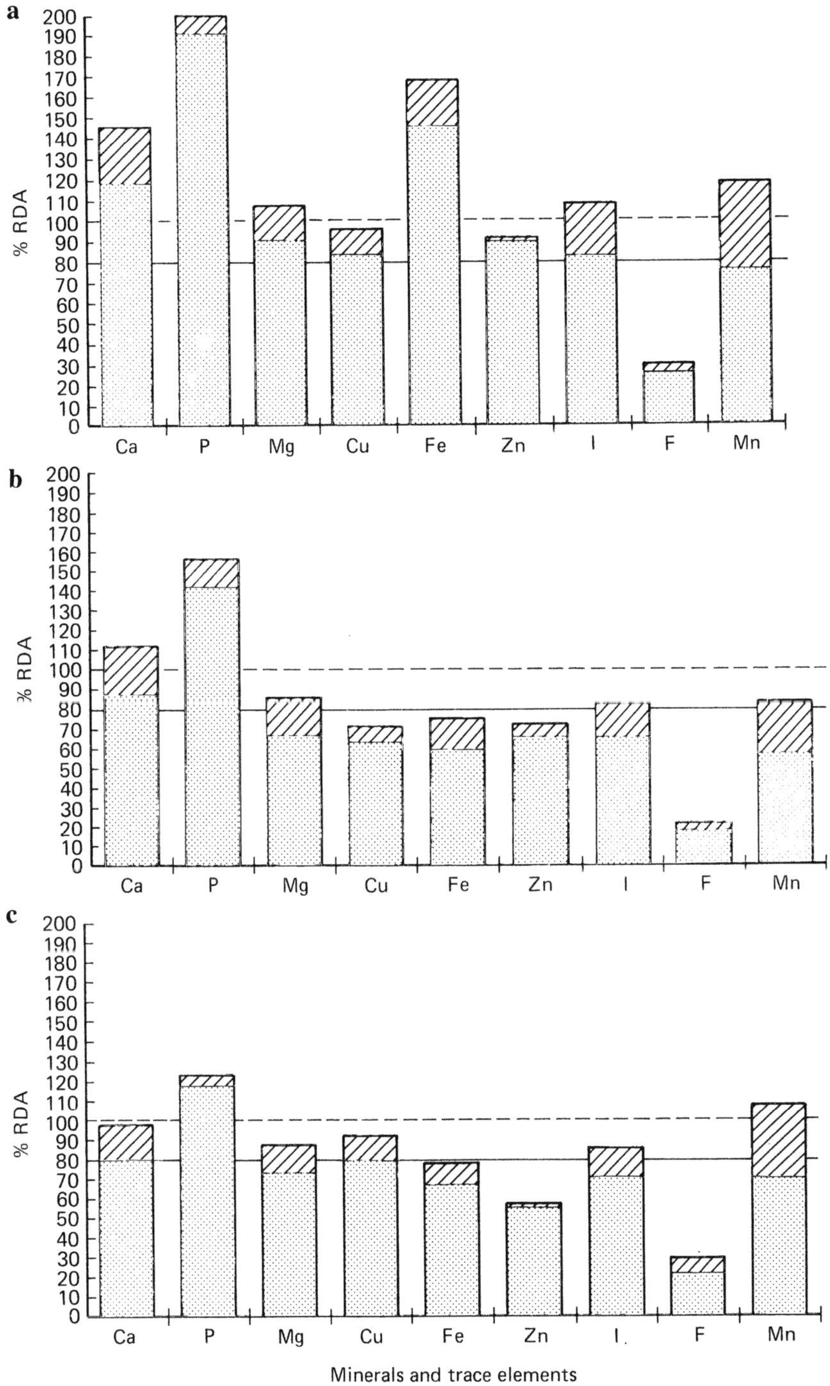

Fig. 4.3a,b,c. Percentages of recommended daily intakes of minerals and trace elements in balanced and observed diets. **a** In men (2700 kcal); **b** in women (2000 kcal); **c** in pregnant women (2500 kcal). *Solid columns,* common diets; *hatched columns,* balanced diets.

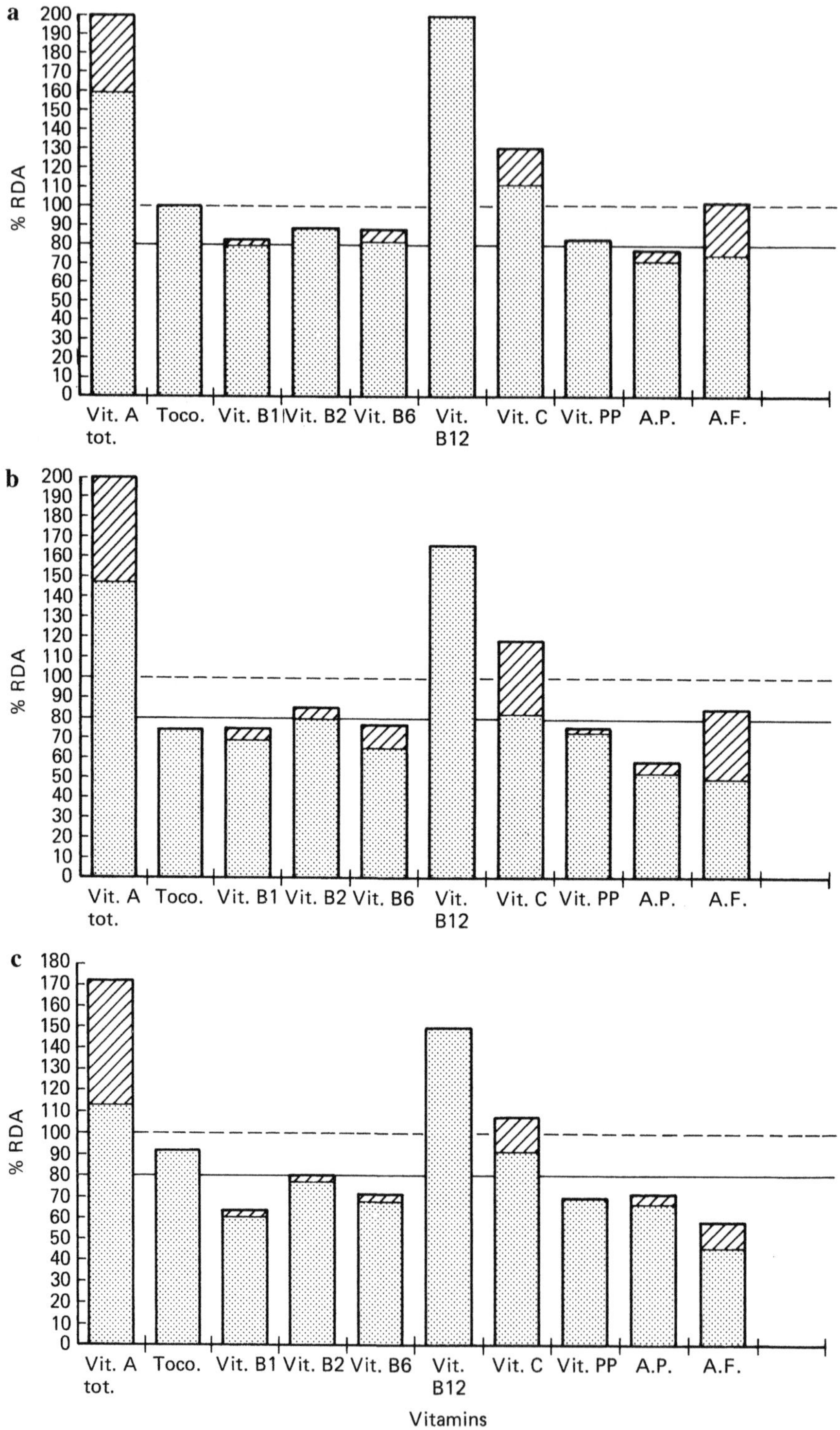

Fig. 4.4a,b,c. Percentages of recommended daily intakes of vitamins in balanced and observed diets. **a** In men (2700 kcal); **b** in women (2000 kcal); **c** in pregnant women (2500 kcal). *Solid columns,* common diets; *hatched columns,* balanced diets.

(1987a) gave the name of Previsional Nutritional Index (PNI) to the list of elements with intakes lower than 80% of that recommended (Table 4.3).

The term "Previsional" is justified in this case because the elements of the PNI list have a good chance of being deficient when supplied by common diets. We know that mineral and vitamin intakes supplied by a balanced diet are slightly higher than those supplied by common diets. The PNI should also help epidemiologists to determine their working programmes.

References

Anderson GH, Peterson RD, Beaton GH (1982) Estimating nutrient deficiencies in a population from dietary record: the use of probability analyses. Nutr Res 2:409–415

Aubree E, Mareschi JP, Magliola C, Mainguy P (1988) Estimation of the average vitamin intake provided by observed and well-balanced diets in adults in France. Nutrition Ernährung 12:695–698

Anonyme (1986) Les Français consomment toujours plus de graisses. Confluences Diététiques, Chiffres OCDE, 43, Paris

Birch GG, Lindley MG (1988) Low calorie products. Elsevier Applied Science, London New York

Couzy F, Aubree E, Magliola C, Mareschi JP (1988) Average mineral and trace element content in daily adjusted menus (DAM) of French adults. J Trace Elem Electrolytes Health Dis 2:79–83

Debry G (1986) Bulletin du Centre de Nutrition Humaine. Centre de Nutrition Humaine, Nancy

Dupin H (1981) Apports nutritionnels conseillés pour la population française. CNRS-CNERNA. Technique et Documentation Lavoisier

Dupin H, Mareschi JP (1984) Habitudes alimentaires – Mode de vie et apports nutritionnels: implications technologiques. Colloque du 14 juin 1984 de la Fondation Française de Nutrition. Fondation Française de Nutrition, Paris, pp 33–44

Fidanza F, Alberti FA, Coli R, Mencarini CA (1984) Food and nutrient consumption of two rural Italian population groups followed for twenty years. Int J Vit Nutr Res 53:91–97

Haeusler L (1985/86) La consommation alimentaire: perception et réalité. Centre de recherche pour l'étude et l'observation des conditions de vie. Consommation 2:3–14

Mareschi JP, Magliola C, Couzy F, Aubree E (1987a) The well-balanced diet and the "at risk" micronutriments: a forecasting nutritional index. Int J Vit Nutr Res 57:79–85

Mareschi JP, Magliola C, Aubree E, Couzy F (1987b) Can the mineral elements intakes recommended for the French population be provided by a balanced diet? Int J Vit Nutr Res 57:225–230

Nelson M (1986) The distribution of nutrient intake within families. Br J Nutr 55:267–277

Paul AA, Southgate DAT (1978) McCance and Widdowson's composition of foods. Elsevier, London New York

Souci SW, Fachmann W, Kraut H (1986) Food composition and nutrition tables 1986/87. Wissenschaftliche Verlagsgesellschaft mbH, Stuttgart

Chapter 5

Biochemical Markers for Micronutrient Status and Their Interpretation

J. Schrijver

Introduction

In nutrition studies a large variety of biochemical markers* are used to evaluate the micronutrient status (Bamji 1981, Christakis 1972, Sauberlich et al. 1974, Sauberlich 1983, 1984a,b, 1986, Wright and Heymsfield 1984). In general, these indicators at appropriate levels are judged to rule out the existence or the threat of a deficiency (Beaton 1986). However, when applying them, a distinction has to be made between apparently healthy individuals, population groups and patients. In the first two groups a series of environmental factors (including recommendations on energy reduction) and physiological factors may influence the nutritional status. For patients the stage of the disease and the accompanying, often long-term, use of drugs may also seriously affect the levels of the indices. In this paper, biochemical indices of vitamin, macro-element and trace element status are discussed in relation to these effects.

The micronutrient status can, in principle, be assessed by determining dietary intake, by clinical examination, and by using clinico-biochemical methods. However, all these approaches suffer from one or more drawbacks.

Dietary Investigation

The average daily dietary intake of nutrients by an individual can be investigated by a food frequency questionnaire, by a 24h recall method and by a food diary (1–7 days). The observed intake can be compared with the recommended daily intake (RDI) and dietary allowance (RDA) for that specific nutrient. However, the accuracy of the recordings varies markedly with the different methods, and the amount and reliability of the information in food composition tables are

*In this paper, the terms "marker", "indicator", "variable", and "index" are used in the same context; the term "parameter" is not used as it is an incorrect indicator term for the tests referred to.

limited. Therefore, investigation of the dietary intake at the individual level will give information on the adequacy of the dietary pattern only.

RDIs/RDAs are recommendations that allow for the individual variation in needs within a population or group. These recommendations can thus be used only to identify a statistical risk of a group, i.e., the risk that a significant percentage of that group will have an insufficient intake of a certain nutrient.

Clinical Examination

For most micronutrients clinical manifestations of deficiencies or overdoses are well documented. However, the specific signs and symptoms of deficiency appear at a relatively late stage, whereas symptoms of early hypovitaminosis are rather vague and rarely specific (e.g., malaise, loss of appetite and weight, and increased irritability). Also with some exceptions such as vitamins A and D, intoxication with micronutrients generally appears in an advanced stage only.

Extended and careful clinical examination, and history-taking, including drug history and dietary habits can provide valuable information. However, laboratory investigations will generally be needed to confirm the diagnosis of a possible depletion of body stores and to demonstrate biochemical abnormalities at a stage before clinical signs and functional impairment become apparent.

Laboratory Analysis

A large variety of laboratory methods for biochemical markers are now available for the assessment of the micronutrient status. These biochemical methods can be classified roughly into two categories:

Direct (i.e., "static") methods, which focus on compositional aspects by measuring the level of a micronutrient or of its metabolites in one or more biological matrices (blood, cerebrospinal fluid, urine, tissues, etc.).

Indirect ("dynamic") methods, which relate to functional aspects by evaluation of the *in vitro* or *in vivo* response of a related biochemical marker (e.g., enzyme activity) or physiological function (immunity, organ function, whole body function) (Solomons and Allen 1983).

The micronutrient status is the net result of the level of intake through the diet, the absorption via the alimentary canal, the uptake by tissues for storage and conversion to its metabolically active form or forms and degradation and excretion rates.

In a deprived state there may be an alteration in possible body stores, in absorption, or in homeostasis at the whole body level or at the level of single organs. It is preferable to determine the level of the micronutrient in its functional surroundings and in its metabolically active form (e.g., thiamin diphosphate in red blood cells; selenium and glutathione peroxidase in platelets). However, the levels of the variables may be affected by a multitude of biological and technical factors (e.g., homeostatic regulation of circulating levels occurring when the supply of an essential nutrient becomes limited, as for depletion of body stores before blood levels of a nutrient are decreased: vitamins A and E, Fe, Ca; or adaptation of absorption: vitamin C, Ca, Fe). There may be a strong effect of infection as well.

For these "static" variables, normal or reference ranges are needed for comparison purposes. Cut-off levels are derived from selected individuals in good health and with well-described living and dietary habits. However, this derivation is purely statistical and is not related to a physiological or biological function. A single measurement in an individual may result in an improper classification. Several other problems arise with setting proper cut-off criteria from healthy "reference" populations, and the concept of reference ("normal") values is sometimes overestimated. Problems of this kind have led to the suggestion of supplementation with a physiological dose of the nutrient to confirm (retrospectively) the existence of a deficiency or to supplement a selected group of individuals prior to the measurement of the "normal" levels. In that respect, functional (*in vivo*) tests can be of great value in establishing and optimizing interpretation criteria, or cut-off points for the direct ("static") biochemical indices, and for the evaluation of the various other indicators.

Some adaptive mechanisms and physiological influences are discussed by van den Berg and others in this volume. Furthermore, drug treatment may be a serious factor in the induction of micronutrient deficiency (Basu 1988, Debry 1984, Hathcock and Coon 1978, Ovesen 1979, Roe 1984, Winick 1983).

Although the "static" levels etc. can offer reliable information on the individual's micronutrient status, the bottom line will be the answer to the question whether that individual can preserve his health and perform all his physical functions. Functionality can be considered at the level of the whole organism and at the organ, tissue, and cellular and subcellular levels (physiological activity and enzyme activities). Clinical and biochemical responses to a dose of a nutrient and the assay of a feature of disturbed metabolism (e.g., tissue uptake, metabolic balance, plasma appearance and clearance after an oral load, and the response of specific binding proteins) are valuable methods in the investigation of functional aspects of the nutritional status.

Five important areas of functional competence are likely to be affected by malnutrition: cognitive ability, disease response, reproductive performance, physical activity and work performance, and social/behavioural performance. However, although deficiency of a nutrient such as thiamin may lead to reduced work performance, the latter does not necessarily indicate the former.

A series of functional indices for laboratory investigation of the micronutrient status of an individual is given in Table 5.1. Materials for *in vivo* and *in vitro* investigation, in decreasing order of availability and ease of collection are: nails, hair, saliva, a single portion of urine, blood (whole blood, plasma and the various cell types), 24-h urine, tear fluid, perspiration, breath, fat, skin, muscle, bone marrow, and liver. As yet, for various more-or-less-obvious reasons, most of these have hardly been used and evaluated in the assessment of nutritional status.

For the development of a reliable functional test of the micronutrient status knowledge of the metabolism of that nutrient is a prerequisite. Furthermore, the discriminative power of the test, i.e., its specificity and its sensitivity, determines the accurate judgement of the micronutrient status of an individual. Using functional criteria, the sensitivity of a test indicates the chance of a positive test result when a deficient status is present, whereas the specificity of a test is the chance of a negative test result in case of an adequate status. In the ideal situation both should equal 100%.

Table 5.1. Classification of functional indices of nutritional status (modified from Solomons and Allen 1983)

Index	Nutrient
In vitro tests of in vivo function	
Capillary fragility	C
Collagen accumulation in implant sponge	Zn, C (?)
Elastin and collagen cross-linking	B6
Enzyme stimulation	B1, B2, B6
Red blood cell fragility	E, Se
Experimental wound healing	C, Zn
[^{14}C]-formate incorporation in serine	Folic acid
Lipid peroxidation (breath ethane/pentane)	E, Se
Prothrombin time	K
Platelet aggregation	E, Zn
Sperm count	Zn
d-Uridine suppression	B12, Folic acid
Tensile strength of skin	Cu
Indusced responses in vivo	
Histidine loading	Folic acid
Tryptophan loading	B6
Valine loading	B12
Fe or Co radioisotope absorption	Fe
[^{14}C]-histidine or [^{14}C]-serine ($^{14}CO_2$ breath)	Folic acid
Hydantoin propionic acid excretion following histidine loading	Mo
Increasing circulating methionine following sodium bisulphite loading	Mo
Post-glucose plasma chromium response	Cr
Post-glucose urine chromium response	Cr
Relative dose response	A
GSH-Px increase in platelets after Se loading	Se
^{75}Se uptake by red blood cells	Se
^{65}Zn uptake by red blood cells	Zn
Thyroid radioiodine uptake	I
Xanthine and sulphur metabolites after purine and sulphur amino acid loading	Mo
Glucose loading followed by exercise	B1
Oxygen/ozone loading (lipid peroxidation)	E, Se
Urinary 3-OH-isovaleric acid after leucine loading	Biotin
Spontaneous in vivo response	
Abducens (VI cranial nerve) function	B1
Central scotoma	A
Colour discrimination	A
Cumulative heart rate	Fe
Dark adaptation/night blindness	A, Zn
Host defence	B6
Mental performance	B6 (?), Zn
Nerve function	B1, B12
Neutrophilic hypersegmentation	Folic acid
Olfactory acuity	A, B12, Zn
Physical performance/endurance	B1, B2, B6, Fe
Taste acuity	A, Zn
Vasopressor response	C
VO_2 (sub)max	Fe
Response of individual or population	
Birth weight	Folic acid, B6
Other	
Conjunctival impression cytology	A

The Group of European Nutritionists (GEN; Fidanza 1986) and the IFCC Expert Panel on the Laboratory Assessment of Nutritional Status (EPLANS) are both involved in the standardization of biochemical indicators of nutritional status. Furthermore, several IUNS committees are studying that aspect: Committee I/12 on "Biochemical methods for the assessment of vitamin status", Committee IV/5 on "Trace elements in human nutrition", and Committee IV/6 on "Functional consequences of vitamin deficiencies". However, the development and the evaluation of the reliability of functional tests of micronutrient status still lack such concerted actions.

When studying deficiencies, one considers a primary deficiency to result from inadequate intake by the normal, healthy, non-pregnant individual, from aberrant or alternative dietary habits or because of increased requirements, such as in the case of pregnancy, lactation, or common diseases. A secondary deficiency arises when the intake is apparently sufficient but inadequate amounts of the micronutrients become available to the body as a result of disturbed utilization, i.e., organic defects (e.g., malabsorption), metabolic diseases, extensive use of medicines, poisoning and intoxication.

Vitamins

Small variations in the daily intake are hardly reflected by the micronutrient levels in blood. However, urinary excretion of B vitamins and of vitamin C more-or-less reflects recent vitamin intake and is therefore of limited value in the assessment of the vitamin status in individuals. It can be used in population studies, for comparison purposes with respect to average intake (Sauberlich 1978, 1984a, Selhub and Rosenberg 1984). Some B vitamins are more-or-less equally distributed between plasma and red blood cells, such as vitamin B6, biotin, niacin, pantothenic acid and vitamin C. Thiamin, riboflavin and folic acid are mainly present in red blood cells, and vitamin B12 and fat-soluble vitamins almost exclusively in plasma (Friedrich 1988). In some cases, such as for folic acid and PLP, determination in plasma is considered more sensitive or specific in spite of the higher content in red blood cells. For some other micronutrients the levels are much higher in blood cells (e.g., vitamin C in leucocytes) than in plasma, more-or-less reflecting body stores or cellular function. With respect to fat-soluble vitamins (especially retinol), plasma levels do not reflect body stores (e.g., liver vitamin A).

Although holo-ETK, -EGR, -EAST, and -EALT levels change as a result of vitamin excess and deficiency, non-nutritional factors, such as drug use and disease may also influence the enzyme levels. The apo-ETK seems more labile both *in vivo* and *in vitro* than the EAST- and EGR-apoenzymes.

With respect to the following sections, reported references values and supposed marginal and deficient levels of a series of biochemical markers of the micronutrient status are summarized in Tables 5.2 and 5.3.

Table 5.2. Observed values for indices of the vitamin status in regular Dutch blood donors (18–65 years, n=207), CIVO-TNO, Zeist, Netherlands, 1986/1987

Vitamin Index	Unit	P 2.5–P 97.5*
Vitamin A (retinol)	μmol/l serum	1.1–3.0
Total carotenoids	μmol/l serum	0.7–4.1
Vitamin E (alpha-tocopherol)	μmol/l serum	15.6–43.8
25-OH-vitamin D	nmol/l serum	19–126
Vitamin B1 (total thiamin)	nmol/l blood	87–168
Transketolase (ETK basal)	U[a]/mmol Hb[b]	7.2–16.3
ETK total	U/mmol Hb	8.1–17.8
Transketolase activation (ETK-AC)	U/U	0.99–1.22
Vitamin B2 (FAD)	nmol/l blood	200–360
Glutathione reductase (EGR, basal)	U/mmol Hb	70–156
EGR total	U/mmol Hb	88–155
EGR activation (EGR-AC)	U/U	0.92–1.26
Vitamin B6 (PLP)	nmol/l blood	35–108
Vitamin B6 (PLP)	nmol/l plasma	18–109
Vitamin B6 (PLP)	nmol/l RBC	31–153
Transaminase (EAST basal)	U/mmol Hb	41–94
EAST total	U/mmol Hb	56–126
EAST activation (EAST-AC)	U/U	1.06–1.72
Vitamin B12 (cyanocobalamin)	pmol/l serum	138–565
Folic acid (5-Me-THF)	nmol/l serum	5.0–19.2
Vitamin C	μmol/l blood	15–69
Biotin	nmol/l blood	0.7–3.7
Niacin	μmol/l blood	35–62
Pantothenic acid	μmol/l blood	0.7–2.1

*percentiles of the range of values observed for men and women.
[a] U, unit of enzyme activity (μmol of substrate converted/min at 37°C).
[b] Hb, haemoglobin; RBC, red blood cells.

Fat-soluble Vitamins

Vitamin A

In addition to its function in vision, vitamin A plays an important but incompletely understood role in the differentiation and maintenance of epithelial membranes and in spermatogenesis, bone growth, and mucopolysaccharide synthesis. Provitamin A carotenoids (e.g. ß-carotene, mol.wt 536.9) serve as precursors of vitamin A, conversion taking place primarily in the small intestine. Together with other carotenoids they contribute to the colour of tissues, e.g., skin. Vitamin A generally comprises retinol (mol.wt 286.5), retinal, and retinoic acid. Vitamin A is mainly stored in the liver as retinyl fatty acid esters, the storage pool increasing with age; carotenoids are stored to a much lesser extent and mainly in the adipose tissue. The reserve vitamin A capacity of the normal adult human body, i.e., the amount available before deficiency signs become apparent at severely inadequate dietary intake, is estimated at 1–2 years (Kübler 1988, Friedrich 1988).

Table 5.3. Index levels indicating vitamin deficiency

Index	Matrix	Unit	Marginal level	Deficient level
Retinol	Serum	μmol/l	0.35–0.70	<0.35
	Liver	μg/g	5–20	<5
RDR[a]		%	10–20	>20
Total carotenoids	Serum	μmol/l	0.50–0.70	<0.50
25-OH-vitamin D	Serum	nmol/l	10–20	<10
Alpha-tocopherol	Serum	μmol/l	12–15	<12
Vitamin K1	Serum	nmol/l	(?)[b]	<0.1 (?)
Thiamin (B1)	Blood	nmol/l	70–90	<70
	Urine	μg/24 h	40–100	<40
		μg/g Creat.[c]	27–66	<27
ETK	RBC	U/mmol Hb	5–7	<5
ETK-AC		U/U	1.20–1.25	>1.25
FAD (B2)	Blood	nmol/l	150–200	<150
	Urine	μg/24 h	40–120	<40
		μg/g Creat.	27–80	<27
EGR	RBC	U/mmol Hb	50–70	<50
EGR-AC		U/U	1.20–1.30	>1.30
PLP (B6)	Blood	nmol/l	20–30	<20
	Plasma	nmol/l	10–15	<10
	Urine	μg/24 h	500–800	<500
		μg/g Creat.	200–300	<200
EAST	RBC	U/mmol Hb	40–50	<40
EAST-AC		U/U	1.80–2.20	>2.20
Vitamin C	Blood	μmol/l	12–17	<12
	Plasma	μmol/l	10–15	<10
	WBC[c]	mg/l	80–150	<80
	Urine	mg/24 h	<8	≪8
Vitamin B12	Serum	pmol/l	75–100	<75
5-Me-THF	Serum	nmol/l	2–4	<2
	RBC	nmol/l	300–350	<300
Biotin	Blood	nmol/l	?	<0.5 (?)
	Urine	μg/24 h	?	<20 (?)
Niacin	Blood	μmol/l	?	<30 (?)
	Urine	mg/24 h	?	<5 (?)
Pantothenic acid	Blood	μmol/l	?	<4 (?)
	Urine	mg/24 h	?	<1 (?)

[a] RDR: relative dose response test (% increase of serum retinol after dosage)
[b] ?: not known/uncertain or insufficient data available.
[c] Creat.: creatinine.
[d] WBC: white blood cells.
See also legend to Table 5.2.

The most reliable index of the vitamin A status would be the amount of vitamin A stored in the liver. As this is impractical, the vitamin A status is investigated by analysis of plasma or serum retinol, although its level in serum is kept quite constant by the liver as long as the liver store is not depleted (>20mg/kg liver). At concentrations below 0.70 μmol/l serum (<200 μg/l), the vitamin A status has probably deteriorated and below 0.35 μmol/l (<100 μg/l) deficiency is most probable. However, low levels are found in protein-energy malnutrition (PEM), zinc deficiency and liver disorders, not reflecting retinol deficiency.

Carotenoids in plasma and serum are poor indicators of the vitamin A status. They merely reflect recent dietary intake from green leafy vegetables and fruit and absorption in the gut. Levels <0.7 μmol/l indicate a low (recent) intake of carotenoids.

The levels of retinol-binding protein (RBP; 30–60 mg/l serum) and prealbumin (PA; 100–400 mg/l serum) are of limited value as they are influenced by a number of factors not directly related to vitamin A, such as infections (Olson 1982).

Recently, the analysis of vitamin A in tear fluid has been described. Whether this level has prognostic value is not yet known (Speek et al. 1986).

The dark adaptation test can be used to detect a developing vitamin A deficiency. Night-blindness appears to be the first clinical sign of vitamin A deficiency. However, routine testing is not simple. The conjunctival impression cytology test (based on the presence of Goblet cells) seems to be reliable and valuable in the assessment of the vitamin A status (Natadisastra et al. 1988; see also the contribution by Amédéé-Manesme in this volume).

The relative dose–response test also seems a promising method, although impractical on a large scale. If the increase of the serum retinol level after a single dose of vitamin A is less than 10% the status is normal, at 10%–20% the result is inconclusive, and at >20% a deficiency is present. This method is based on the observation that in a vitamin-A-deprived state RBP synthesis continues in hepatocytes in excess of holo-RBP that is released. When vitamin A becomes available, it binds to RBP and is promptly released into the blood (Mobarham et al. 1981).

Primary deficiency has repeatedly been described in fast growing children, especially in children with infections such as measles, as a result of PEM, and in children on a low-fat diet, resulting in the various stages of xerophthalmia and ultimately blindness. Secondary deficiency may result from malabsorption, lipoproteinaemia, long-term drug use (cholestyramine, neomycin, mineral oil, antacid, cortisone, phenobarbital, caffeine), and alcohol abuse. Clinical signs include degenerative changes in the eyes and skin.

Vitamin D

Vitamin D (mol.wt 384.7; for 25-OHD mol.wt 400.7) is required for the development and maintenance of the skeleton and for a proper absorption and use of calcium. Various hydroxylated metabolites of vitamin D can be demonstrated. In healthy adults the following plasma/serum levels are found: 2–12 nmol/l vitamin D, 20–130 nmol/l 25-hydroxyvitamin D (25-OHD), 2–10 nmol/l 24,25-dihydroxyvitamin D (24,25-diOHD) and 0.04–0.14 nmol/l 1,25-diOHD. The 25-OHD level, which depends on sunlight and is at its highest level in June in Europe, reflects both dietary supply of vitamin D (Whyte 1979) and synthesis by the skin (Devgun et al. 1981). In early stages of vitamin D deficiency its low level may be the sole abnormality (Bouillon and De Laey 1983). The 1,25-diOHD level is regulated within a narrow range, and hardly reflects the supply of vitamin D, but merely kidney function. The level of alkaline phosphatase in the circulation is influenced by various factors and circumstances and is therefore not a reliable indicator of the vitamin D status. Calcium and phosphate levels in serum also are regulated within narrow limits. As their levels may change as a result of various afflictions they cannot be used as reliable indicators for the vitamin D status. Dynamic biochemical methods for the assessment of vitamin D status are not known.

In general, primary deficiency results from lack of sunlight (UV) exposure. Home-bound elderly persons, dark-pigmented people in cold climates and fast-growing children (up to 6 years, e.g., those breast-fed with milk deficient in vitamin D) are the main risk groups. The reserve vitamin D capacity of the normal adult human body, i.e., the amount available before deficiency signs become apparent at severely inadequate dietary intakes and low exposure to UV is estimated at 2–4 months (Kübler 1988, Friedrich 1988). Secondary deficiency has been described in malabsorption, parathyroid disorders and long-term drug use (anticonvulsants, cholestyramine, glutethimide, irritant cathartics, mineral oil, neomycin). Vitamin D deficiency manifests itself as rickets and osteomalacia. Its role in osteoporosis is not yet fully understood (Ziegler 1988).

Vitamin E

Vitamin E refers to a family of lipid-soluble tocopherols and tocotrienols of which d-alpha-tocopherol (mol.wt 430.7) is the most active member. It acts as a physiological antioxidant and exerts a stabilizing action on hormones, enzymes, other vitamins and lipids. It stabilizes cellular and lysosomal membranes and prevents the formation of lipoperoxides. The reserve vitamin E capacity of the normal adult human body (i.e., the amount available before deficiency signs become apparent at severely inadequate dietary intake) is estimated at 6–12 months (Kübler 1988, Friedrich 1988).

Vitamin E status is often assessed by analysis of alpha-tocopherol in plasma and serum (preferably in relation to the lipid level) and in red blood cells in which it is present exclusively within the cell membrane. In adults serum levels of 15–44 μmol/l, which slowly increase with age, are found, with an average of 1.8 mg vitamin E per g lipid. Deficiency is probably present at a level <2 mg/l (<5 μmol/l). The red blood cell haemolysis test, where resistance is measured of red blood cells to oxidative stress in the presence of peroxide and normal values $<10\%$ haemolysis can be expected, has been used as a functional indicator. However, this method has a low discriminative power. Abnormal values are found at serum E <5 mg/l (<12 μmol/l). The PUFA level in the diet has a strong effect on the result of this peroxide haemolysis test.

Primary deficiency is sometimes reported in premature babies and low-birth-weight infants (Huybers et al. 1986) as a result of poor placental transport and limited adipose tissue. Secondary deficiency may occur in chronic malabsorption syndromes, lipoproteinaemia, long-term drug use (mineral oil, cholestyramine, broad-spectrum antibiotics such as neomycin), and high-PUFA parenteral nutrition. Haemolysis is the main clinical sign of vitamin E deficiency.

Vitamin K

Vitamin K (mol.wt 450.7 for vitamin K1) is required for the synthesis by the liver of prothrombin (through carboxylation of glutamate residues) and three other factors of the extrinsic coagulation scheme: factors VII, IX, and X. It plays a role in electron transport along the respiratory chain.

For the analysis of the various vitamin K1 forms (free and epoxides) in serum no definite method is available (reported normal levels range from 0.1 to 2.2

nmol/l or higher, Lambert et al. 1986, Guillaumont et al. 1988). In general, vitamin K status is investigated through the prothrombin test. In this test the time required for clotting (prothrombin time, i.e., conversion of prothrombin into thrombin) is measured (normal values: 11–18 s). However, no change is observed in this time period until the levels of circulating prothrombin drop below 30% or normal. Furthermore, a serious risk of bleeding does not occur even during surgery until the prothrombin content is less than 20%–40% of normal as reflected in a prolongation of prothrombin time by at least 3–4 s. It should be emphasized that vitamin K deficiency is not the only clinical condition that results in a prolonged prothrombin time. Since prothrombin is synthesized by hepatocytes, any condition that causes a defect in hepatocellular function or loss of liver tissue such as severe cirrhosis results in a prolonged prothrombin time that is unresponsive to vitamin K treatment. Similarly, anticoagulants such as coumarins are designed to interfere with vitamin K function and to prolong prothrombin time.

Primary deficiency has been described for situations of insufficient intake and/or synthesis by the gut flora. The reserve vitamin K capacity of the normal adult human body, i.e., the amount available before deficiency signs become apparent at strongly inadequate dietary intake and synthesis by the gut flora, is estimated at 2–6 weeks (Kübler 1988, Friedrich 1988). Secondary deficiency is observed in malabsorption with steatorrhoea, total parenteral nutrition, liver malfunction, and long-term drug use (antibiotics, anticonvulsants, cholestyramine, mineral oil, salicylates, cinchona alkaloids, coumarin, barbiturates). Vitamin K deficiency in haemorrhagic disease of the newborn has not yet been proved.

Water-soluble Vitamins

Thiamin

Thiamin (vitamin B1, mol.wt 337.3 for thiamin chloride HCl) functions as the coenzyme thiamin diphosphate (ThDP) in the metabolism of α-keto acids and 2-keto sugars and as thiamin triphosphate (ThTP) in nervous tissue in a manner not yet fully understood.

At the individual level, urinary excretion of thiamin (critical level 40 μg/24h; 27μg/g creatinine) is not reliable as an indicator of the vitamin B1 status at the individual level because of the strong influence of recent dietary intake. For investigations of populations it corroborates other indicators. The red-blood-cell activity of transketolase (ETK) and its *in vitro* activation by ThDP (ETK-AC; levels >1.25 indicate deficiency) are often used because they can be considered as functional indicators of the thiamin status at the cellular level. However, it should be recognized that the synthesis of transketolase is reduced in thiamin deficiency resulting in normal ETK-AC values at prolonged deficiency. In that situation, in particular, the concentration of thiamin (deficiency at a level below about 70 nmol/l) or ThDP in whole blood or red blood cells can serve as a more conclusive or additional indicator. The very low concentration of thiamin in plasma and serum (<10 nmol/l) is not suitable as a marker of the thiamin status. As is shown in Table 5.4 for a normal population (Dutch blood donors), significant correlations are found between the basal ETK and total ETK activities and the concentration of total thiamin in whole blood (Schrijver et al. 1982, 1985).

Table 5.4. Correlation matrix of indices of the thiamin status (Pearson correlation coefficients)

Index	Thiamin	Basal ETK	Total ETK
Total thiamin	1.0000		
Basal ETK	0.4201*	1.0000	
Total ETK	0.3814*	0.9721*	1.0000
ETK-AC	−0.2898	−0.3997*	−0.1806

* Significant at P <0.001 (df=205); blood samples of Dutch blood donors (men and women, 18–65 yr) were used for the investigation of the indices of the thiamin status.

The primary deficiency of thiamin results in the various classical forms of beri beri. The reserve thiamin capacity of the normal adult human body, the amount available before deficiency signs become apparent at severely inadequate dietary intake, is estimated at 1–2 weeks (Kübler 1988, Friedrich 1988). Secondary deficiency in developed countries is mainly found in cases of chronic alcoholism and alcohol intoxication (Wernicke–Korsakoff syndrome). Sometimes the presence of thiaminase in some food products, for example, raw fish and the long-term use of drugs such as diuretics, antacids and fluoruracil are risk factors. Renal dialysis (Boeschoten et al. 1988) and parenteral feeding as well as a thiamin-responsive inborn error of metabolism may result in thiamin deficiency. In general, early deficiency signs are diffuse and vague and include mental disturbances such as depression, memory loss and so on, and muscular weakness.

Riboflavin

Riboflavin (mol.wt 376.4) functions primarily as the reactive portion of FMN, mol.wt 456.4, and FAD, mol.wt 785.6, the prosthetic groups of about 60 flavoproteins, enzymes involved in a variety of biological oxidations.

Riboflavin status can be assessed by analysis of FAD, FMN and riboflavin itself in whole blood, plasma, serum or red blood cells (Speek et al. 1982). The red-blood-cell glutathione reductase (EGR) activity and its *in vitro* stimulation by FAD (EGR-AC; deficiency at levels >1.30) are functional markers at the cellular level. The level of riboflavin in urine (critical level 40 μg/24 h; 27 μg/g creatinine) is a less reliable marker because of the strong influence of recent dietary intake. Table 5.5 shows significant correlations between the basal and total EGR activities and FAD in whole blood as observed for healthy adults.

Table 5.5. Correlation matrix of indices of the riboflavin status (Pearson correlation coefficients)

Index	FAD	Basal EGR	Total EGR
FAD	1.0000		
Basal EGR	0.5065*	1.0000	
Total EGR	0.4725*	0.9076*	1.0000
EGR-AC	−0.2690	−0.5800*	−0.2012

* Significant at P <0.001 (df=204); blood samples of Dutch blood donors (men and women, 18–65 yr) were used for the investigation of the indices of the riboflavin status.

Primary deficiency in developed countries is not known. The reserve riboflavin capacity of the normal adult human body, the amount available before deficiency signs become apparent at severely inadequate dietary intake, is estimated at 2–6 weeks (Kübler 1988, Friedrich 1988).

Secondary deficiency can be the result of malabsorption, alcohol abuse, boric acid overdose or long-term drug use (e.g., probenecid, chlorpromazine, phenothiazine). Deficiency signs in a more-or-less-advanced stage are a sore throat, hyperaemia, angular stomatitis, cheilosis, seborrheic dermatitis and low haemoglobin. With serious deficiency a normochromic, normocytic anaemia and reticulopenia are observed.

Vitamin B6

Vitamin B6 (pyridoxine, pyridoxal and pyridoxamine) in the form of pyridoxal 5′-phosphate (PLP, mol.wt 247.1) functions as the prosthetic group of enzymes involved in transamination and other reactions in the metabolism of fat and carbohydrates. The assessment of the vitamin B6 status has often been investigated by determining the PLP level in plasma or whole blood (Schrijver et al. 1981). Blood plasma levels below 20 (10) nmol/l are considered to be deficient. Hormonal effects may strongly affect the PLP level especially in plasma, but do not indicate a deficiency (van den Berg, this volume). Additionally, the red-blood-cell aspartate transaminase (EAST) and alanine transaminase (EALT) activities and their *in vitro* stimulation by PLP (EAST-AC and EALT-AC, respectively) are used as indicators of the vitamin B6 status. EAST-AC values indicating a vitamin B6 deficiency seem to depend strongly on the actual procedures used for the kinetic analysis of transaminase (EAST-AC values >2.20 are reported as indicative of vitamin B6 deficiency). The activity of EALT is much lower and less stable *in vitro* than is EAST. Table 5.6 shows significant correlations of whole blood and red blood cell PLP with the basal and total EAST activities for healthy adults. The lack of such a correlation for plasma PLP renders it questionable as a marker of vitamin B6 available to tissues.

The excretion of 4-pyridoxic acid, the main metabolite of vitamin B6 in urine (deficiency at <0.5 mg/24h), is not a suitable marker at the individual level due to the strong influence of recent dietary intake. Oral tryptophan loading (Schrijver et al. 1984) and methionine loading have been used as functional tests of the vitamin B6 status. However, the results of these tests may be seriously affected by other factors not related to vitamin B6 (e.g., oestrogen-containing oral contraceptives).

Primary deficiency has been described for sucklings. Insufficient intake of niacin and riboflavin contribute to vitamin B6 deficiency. The reserve vitamin B6 capacity of the normal adult human body (i.e., the amount available before deficiency signs become apparent at severely inadequate dietary intake) is estimated at 2–6 weeks (Kübler 1988, Friedrich 1988). Secondary deficiency may result from vitamin B6 antagonists (e.g., isoniazid), long-term drug use (chlorpromazine, hydralazine, cycloserine, penicillamine, alpha-methyldopa), inborn errors of metabolism, malabsorption, alcohol abuse, uraemia or liver disease. Deficiency signs include: convulsions, vomiting and weight loss in infants, and EEG abnormalities, convulsions, depression and confusion in adults. An advanced deficiency results in hypochromic anaemia, a decrease of lymphocyte count, and possibly a normocytic, microcytic, or sideroblastic anaemia. Vitamin

Table 5.6. Correlation matrix of indices of the vitamin B6 status (Pearson correlation coefficients)

Index	PLP			EAST	
	Blood	Plasma	RBC	Basal	Total
PLP					
Blood	1.0000				
Plasma	0.6889*	1.0000			
RBC	0.7561*	0.0621	1.0000		
Basal EAST	0.4120*	0.1823	0.3912*	1.0000	
Total EAST	0.2993	0.0387	0.3581*	0.7981*	1.0000
EAST-AC	−0.2222	−0.2571	−0.0837	−0.3800*	−0.2382

* Significant at P <0.001 (df=205); blood donors (men and women, 18–65 yr) were used for the investigation of the indices of the vitamin B6 status (RBC=red blood cells).

B6 deficiency is alleged to be involved in some physical and mental disturbances which appear to accompany ageing.

Vitamin B12

In man, vitamin B12 (mol.wt 1355.4) is involved in two reactions: transmethylation of homocysteine to methionine by 5-methyl-tetrahydrofolic acid and isomerization of methylmalonyl-CoA to succinyl-CoA.

The vitamin B12 status is best reflected by the store of the vitamin in the liver. It is obvious that this is an impractical approach in normal individuals. Therefore, levels in plasma, serum and red blood cells are determined, although these markers are not very discriminative. The vitamin is extracted by boiling at low pH in the presence of cyanide, which frees the vitamin from its binding protein and converts the various coenzyme forms into cyanocobalamin. If purified intrinsic factor is used in the vitamin assay, the method is specific for biologically active cobalamins since inactive B12 analogues do not react with purified intrinsic factor in the assay. In general, vitamin B12 levels below 100 ng/l serum are considered as insufficient with a concomitant increased urinary excretion of methylmalonate. Reliable methods to gain more insight into the vitamin B12 status, especially to discriminate between vitamin B12 deficiency and folic acid deficiency (although not practical in routine analysis; see also the next section), are the deoxyuridine suppression test (Das and Herbert 1978, Das et al. 1978), the separate analysis of the two biologically most active forms adenosylcobalamin and methylcobalamin, the histidine loading test (urinary excretion of formiminoglutamic acid (FIGLU) before and after an oral histidine load) and the Schilling test. In the latter test an oral dose of ^{57}Co-vitamin B12 is given. At a normal vitamin B12 status more than 8% of the label is recovered in the urine, while at deficiency less than 7% and very frequently even as little as 3% is recovered. In this way one can differentiate between vitamin B12 deficiency and deficiency of the intrinsic factor. Some interpretation criteria are summarized in Table 5.7.

Primary deficiency is rare in Western Europe; low to very low plasma levels have been reported for vegans (strict vegetarians excluding from their diet all food of animal origin). The reserve vitamin B12 capacity of the normal adult

Table 5.7. Cut-off levels of vitamin B12 and folic acid status.

Index	Deficient levels	Low levels	Acceptable levels
Vitamin B12			
Serum level (μg/l)	<100	100–150	>150
Methylmalonic acid (mg/24 h)	>25		<12
Folic acid			
Serum level (μg/l)	<3.0	3.0–5.9	>6.0
Red blood cell level (μg/l)	<140	140–159	>160
FIGLU urinary excretion (mg/24 h)	>7.0		0–7.0
FIGLU after histidine load	>50		5–20

human body (i.e., the amount available before deficiency signs become apparent at severely inadequate dietary intake) is estimated at 3–5 years (Kübler 1988, Friedrich 1988). Secondary deficiency may result from disturbed absorption and utilization, e.g., in chronic atrophic gastritis and (partial) gastrectomy, coeliac disease and sprue, in terminal ileitis (main item: disturbed secretion of intrinsic factor) as a result of fish tapeworm infestation, in inborn errors of metabolism, in long-term drug use (antibiotics, *p*-aminosalicylic acid, biguanides, colchicine, cimetidine, cholestyramine, oestrogens, guanides, nitrous oxide, potassium chloride, trifluoperazine), and as a result of high fibre intake. The ultimate deficiency state is macrocytic (MCV >110 fl) normochromic anaemia (pernicious anaemia).

Folic Acid

Folic acid in its reduced forms (especially 5-methyl-tetrahydrofolic acid, 5-Me-THF, mol.wt 459.5) serves as a prosthetic group in enzymes involved in 1-carbon transfer. To assess its status in individuals, the levels of 5-Me-THF in plasma and red blood cells are generally used. For most clinical situations serum folic acid is adequate, since low body stores are reflected by low levels. In some cases, for example at increased cellular or tissue turnover (e.g., pregnancy, malignancies, haemolytic anaemia, etc.) serum folic acid may not be a good index of the folic acid status. Red-blood-cell folic acid originates at the time of development of the red blood cell and hence is a better index of folic acid stores. Folic acid deficiency is probable at serum levels below 1 μg/l (2 nmol/l) and red blood cell levels below 100 μg/l (200 nmol/l).

Urinary folic acid is not related to intake as a result of synthesis by the gut flora (biologically not available to man). High levels of folic acid are present in bile, indicating an extensive enterohepatic circulation.

Histidine is normally converted into glutamate through a pathway that is folic acid-dependent. At deficiency, histidine loading (2–15 mg) results in increased urinary FIGLU and urocanic acid excretion (Sauberlich et al. 1974). However, this test is less specific, especially during pregnancy. High urinary levels of adenine precursors (e.g. amino-imidazole carboxamide) are found in folic acid deficiency. Folic acid and vitamin B12 deficiency can be discriminated by means of the deoxyuridine suppression test. A cell system is able to synthesize deoxythymidine monophosphate (dTMP) *de novo* from added deoxyuridine, a folic acid-dependent pathway, thereby diminishing, or competing for, incorporation of radioactive thymidine into cellular DNA. In folic acid deficiency, suppression by deoxyuridine of [^{3}H]thymidine incorporation is diminished. The method uses aspirates of bone marrow or peripheral lymphocytes (Das and Herbert 1978, Das et al. 1978). Another functional test of folic acid status uses lymphocyte incorporation of [^{14}C]formate into serine or methionine in the presence of excess glycine or homocysteine (Ellegaard and Esmann 1970). The method is based on the presence of tetrahydrofolic acid coenzymes; in a state of deficiency a decrease in the incorporation of radioactive formate is observed. Haematological indices (e.g., neutrophilic hypersegmentation and MCV) are usually not specific enough for routine analyses of the folic acid status. Some interpretation levels of the folic acid status are summarized in Table 5.7.

Primary deficiency has been described in pregnancy. The reserve folic acid capacity of the normal adult human body (i.e., the amount available before deficiency signs become apparent with severely inadequate dietary intake) is estimated at 2–4 months (Kübler 1988, Friedrich 1988). Secondary deficiency may result from haemolytic anaemia, malabsorption (sprue, Crohn's disease), metabolic disorders, alcohol abuse, and long-term drug use (e.g., methotrexate, aminopterine, hydantoins, barbiturates, antimalaria drugs). Vitamin B12 deficiency results in lower folic acid levels in red blood cells and higher serum folic acid levels, while zinc deficiency decreases the absorption of dietary polyglutamates of folic acid. The major clinical manifestation of folic acid deficiency is macrocytic (MCV >110 fl), megaloblastic anaemia. The main risk groups are alcoholics, individuals with a vitamin B12 deficiency and women on long-term oral contraceptive use before frequent pregnancies.

Niacin

Niacin (nicotinic acid, mol.wt 123.1; nicotinamide, mol.wt 122.1) is a component of NAD and NADP. Both coenzymes are involved in oxidation-reduction reactions. The major method used for the assessment of the niacin status in individuals is the determination of its metabolites N^{1}-methylnicotinamide (normal levels in adults >1.6 mg/g creatinine; deficiency probable at levels <0.5 mg/g) and/or N^{1}-methyl-2-pyridone-5-carboxylamide (2-pyridone) in urine. However, both levels are age- and pregnancy-dependent, so that the N^{1}-methylnicotinamide/2-pyridone ratio would be a better indicator (ratio for all ages 1.0–4.0; deficiency certain at a ratio below 0.5). The N^{1}-methyl derivative continues to decline with progressive deficiency when 2-pyridone is already essentially absent (Sauberlich et al. 1974). It is not certain whether blood levels of niacin are reliable indicators of its status. With the microbiological method (*Lactobacillus plantarum*), levels of 50±7 μmol/l blood have been found in healthy Dutch blood donors (18–65 years of age) and slightly (5%–10%) lower levels for women; a small increase (5%) is found during oral contraceptive use. Levels in red blood cells and plasma are comparable.

Primary deficiency (pellagra) is unknown in affluent societies but has been described for populations in which maize (in which niacin is largely unavailable) is the main component of the diet. The reserve niacin capacity of the normal adult human body (i.e., the amount available before deficiency signs become apparent at severely inadequate dietary intake) is estimated at 2–6 weeks (Kübler 1988, Friedrich 1988). Secondary deficiency may be the result of malabsorption, metabolic disorders and diseases related to abnormal metabolism of niacin or of coenzymes containing niacin, some parasitic infections (deranged absorption), and long-term use of isoniazid (used for treatment of tuberculosis).

Pantothenic Acid

As part of coenzyme A (CoA), d-pantothenic acid (mol.wt 219.2) plays a key role in acyl-group activation reactions. No conclusive tests are available for the assessment of the pantothenic acid status. Tests should evaluate the CoA adequacy in the body. Since urinary output of pantothenic acid is directly proportional to dietary intake, present assessment methods rely mainly on this index. Urinary excretion of <1 mg/24 h is considered abnormally low. Suspicion of inadequate intake is further supported if whole blood values are well below 0.5 μmol/l. Normal levels by microbiological assay (*L. plantarum*) in healthy Dutch adults (regular blood donors, 18–65 years of age) are 1.3±0.5 μmol/l whole blood, slowly increasing with age. No influence of gender is seen, but slightly higher levels are observed in oral contraceptive users (mean 1.3 μmol/l blood versus 1.1 μmol/l blood for non-users). Levels in red blood cells and plasma are comparable. However, pantothenic acid levels obtained through microbiological methods by different laboratories sometimes do not match.

Pantothenic acid is widely distributed in food ("pantothen", from every side), so that only in cases of grossly inadequate diets might deficiency develop. Deficiency has been suspected in the "burning feet syndrome". The reserve pantothenic acid capacity of the normal adult human body (i.e., the amount available before deficiency signs become apparent at severely inadequate dietary intake) is not known. Secondary deficiency might result from chronic severe alcoholism, but probably as a result of dietary deficiency. In conditions of diseased mucosa of the alimentary canal the utilization of pantothenic acid can be disturbed. Experimental deficiency signs include irascibility, postural (orthostatic) hypotension, rapid heart rate on exertion, epigastric distress with anorexia and constipation, numbness and tingling of the hands and feet, hyperactive deep tendon reflexes, and weakness of finger extensor muscles.

Biotin

D(+)biotin (mol.wt 244.3) is involved as a prosthetic group in carboxylase (CO_2 transfer) reactions of pyruvate, propionyl, β-methylcrotonyl-CoA and acetyl-CoA. In general, the biotin status is investigated by determining its level in whole blood or plasma. Levels in red blood cells and plasma seem to be similar. By microbiological assay (*L. plantarum*) in whole blood of healthy adults (Dutch blood donors, 18–65 years of age) levels of 0.7–3.7 nmol/l (0.2–0.9 μg/l) were found, with no effect of age, gender or oral contraceptive use. Deficiency is probable at levels <0.5 nmol/l (<0.1 μg/l). Levels in red blood cells and plasma are comparable. However, biotin levels obtained through microbiological methods by different laboratories sometimes do not match. Normal levels of urinary excretion are in the range of 30–60 μg/24 h.

Functional indices of the biotin status are the carboxylase activities in leucocytes, lymphocytes or fibroblasts in the presence and absence of added biotin. In biotin deficiency or inborn errors of metabolism the urinary excretion of certain organic acids that serve as precursors of the various carboxylases is increased.

As the intestinal microflora contributes strongly to the body pool of available biotin, primary deficiency of the vitamin is very rare. It has been described in cases of frequent consumption of raw egg-white ("egg-white injury"). The reserve biotin capacity of the normal adult human body (i.e. the amount available before deficiency signs become apparent due to inadequate dietary intake and insufficient synthesis by the gut flora) is not known. Secondary deficiency may be present in chronic alcoholism, biotin-responsive inborn errors of metabolism, short-bowel syndrome and in situations of reduced gastric secretion. The main deficiency signs include anorexia, nausea, vomiting, glossitis, pallor, depression, dry and scaly dermatitis (mainly observed in experimental deficiency), and seborrheic dermatitis in infants.

Vitamin C

Ascorbic acid (mol.wt 176.1) is one of the vitamins for which precise biochemical functions remain unknown. It seems to be involved in oxidation and reduction reactions, tyrosine and collagen metabolism, hydroxylation, steroid synthesis, iron utilization and wound healing.

For the investigation of the vitamin C status of the individual (the estimated normal whole body store is about 1–1.5 g), the vitamin is analysed in whole blood, red blood cells, plasma or leucocytes. The analysis of vitamin C may be hampered by the relative instability of this vitamin during normal collection, handling and storage of the biological samples (Speek et al. 1984). Leucocytic vitamin C is a direct reflection of body stores: normal levels at >150 mg/l (>20 μg/10^8 cells; >850 μmol/l), and at >3 mg/l serum (>17 μmol/l). A marginal vitamin C status would be present at 80–150 mg/l leucocytes (450–850 μmol/l) and 2.0–2.9 mg/l serum (11–17 μmol/l). Deficiency is almost certain for vitamin C levels <80 mg/l leucocytes (<450 μmol/l), <2.0 mg/l serum (<11 μmol/l), and <3.0 mg/l whole blood (<17 μmol/l). The urinary level of vitamin C is strongly influenced by recent intake and therefore of limited value in the assessment of the vitamin C status of an individual. Whether the ascorbic acid/dehydro-ascorbic acid ratio in the various matrices is a valuable index of the vitamin C status (mainly in relation to its antioxidant function) needs further investigation.

The classical dietary deficiency manifestation is scurvy. The reserve vitamin C capacity of the normal adult human body, the amount available before deficiency signs become apparent with severely inadequate dietary intake, is estimated at 2–6 weeks (Kübler 1988, Friedrich 1988). Secondary deficiency may result from chronic alcoholism, heavy smoking, acute stress due to surgery or trauma, and long-term drug use (oestrogens, salicylates, steroidal agents, such as prednisone, tetracycline, barbiturates, hydantoins, ether and some other anaesthetics and aminopyrine, paraldehyde and calcitonin). Early deficiency signs are reduced urinary excretion of vitamin C, a decrease of body weight, greater susceptibility to infection and stress, lassitude, fatigue and muscular pain. More severe deficiency causes tachycardia, dyspnoea, and ultimately the potentially fatal scurvy with anaemia, perifollicular bleeding, subperiostal bleeding in children, and the bayonet-rib syndrome in infants.

Macroelements and Trace Elements

In contrast to the assessment of the nutritional status of macroelements that of the trace elements is still in its infancy. Of all the essential macroelements, only four are usually measured in surveys: calcium, phosphorus, magnesium and iron. Of these, calcium and iron intakes are often significantly lower than required. In contrast to most vitamins, minerals and trace elements have, in general, a limited range for which the related biological function is optimal. Below that range deficiency exists, above it a toxic state will develop.

Over the past decades, as analytical detection limits went down, the list of trace elements claimed to be essential for man or animals has grown steadily (Golden 1983, Hambridge and Mauer 1978, Lindeman 1984, Nielsen 1980, Prasad 1982, Prasad and Oberleas 1976, Rennert 1984, Underwood 1977, WHO 1973). The trace elements currently accepted as essential in animals or man are, in increasing order of atomic weight, F, Si, V, Cr, Mn, Co, Ni, Cu, Zn, As, Se, Mo and I. With the exception of two of these (fluorine and iodine), all are trace metals. The essentiality of nickel, vanadium, arsenic, and silicon for humans has been hypothesized but not yet demonstrated. The list of macroelements comprises: H, C, N, O, Na, Mg, P, S, Cl, K, Ca and Fe. Several other elements such as aluminium, barium, boron, bromine, cadmium, germanium, lead, mercury, rubidium, strontium, tellurium, titanium and zirconium have been demonstrated in the human body in varying amounts. Whether or not these elements are essential to human (and animal) life is not yet known. Up to now these elements appear physiologically unimportant and/or have mild or severe toxicological properties.

Some serum and urinary excretion levels as observed in Dutch healthy adults are given in Tables 5.8 and 5.9.

Macroelements

Calcium

The body of an adult man contains about 1.2 kg of calcium (mol.wt 40.08) of which up to 99% is deposited within the bones, mainly as hydroxyapatite, thus constituting an important structural component of the skeleton. The remainder is present in teeth, soft tissues, blood and extracellular fluid.

For an accurate assessment of the calcium status, the calcium content of bone should be estimated. This can be done by various methods, such as histomorphometry (bone biopsy), photon absorptiometry, neutron activation, and computer tomography. However, these methods are impractical in routine screening. Therefore, in spite of its drawbacks, analysis of biological fluids is normally chosen. The level of ionized calcium in plasma is regulated within a narrow range (1.15–1.30 mmol/l); the total concentration of calcium in plasma is 2.20–2.55 mmol/l (88–102 mg/l). Urinary excretion of calcium is dependent on dietary intake. In general, urinary excretion is less than 7.5 mmol/24h (<300 mg/24h) in men and less than 6.25 mmol/24h (<250 mg/24h) in women. In adults the calcium/creatinine ratio in urine is less than 0.40 on a molar or <0.14 on a weight basis.

Table 5.8. Observed Dutch values for indices of the macroelement/microelement status CIVO-TNO, Zeist, Netherlands, 1986/1987

Index	Unit	P 2.5–P 97.5
Aluminium (Al)	μmol/l serum	0.5–7.5
	μmol/24 h urine	0.2–3.3
Arsenic (As)	nmol/l blood	0–400
	nmol/l urine	0–400
	nmol/g hair	0–20
Calcium (Ca)	mmol/l serum	2.25–2.85
	mmol/24 h urine	2.5–8.0
Chromium (Cr)	nmol/l serum	0–40
	nmol/24 h urine	10–80
Cobalt (Co)	nmol/l serum	30–50
Copper (Cu)	μmol/l serum	12–29
	μmol/24 h urine	0.3–1.6
Fluorine (F)	μg/l serum	0–50
	mg/g creatinine	0–2.0
Gold (Au)	μmol/l serum	1.6–20
Iron (Fe)	μmol/l serum	9–31
	μmol/24 h urine	0.7–2.7
Lithium (Li)	mmol/l serum	0.3–1.0
Magnesium (Mg)	mmol/l serum	0.75–1.50
	mmol/24 h urine	0.6–12
Manganese (Mn)	nmol/l serum	10–180
	nmol/24 h urine	50–275
Molybdenum (Mo)	μmol/l serum	0.1–1.6
Nickel (Ni)	nmol/l serum	0–75
	nmol/l urine	180–1000
Quicksilver (Hg)	nmol/l blood	0–25
	nmol/l urine	0–100
Selenium (Se)	μmol/l serum	0.8–1.8
	μmol/l urine	0.06–0.57
Vanadium (V)	μmol/l urine	0–1.0
Zinc (Zn)	μmol/l serum	11.5–23.5
	μmol/24 h urine	4.5–14

* Percentiles of the range of values observed for men and women.

Calcium deficiency is very rare. Primary deficiency has been described in young fast-growing children, with symptoms comparable to rickets. The question remaining to be answered is whether calcium is involved in osteoporosis. A high protein intake increases the calcium requirement. A high magnesium intake may cause hypocalcaemia; vitamin D deficiency results in a disturbed absorption of calcium in the small intestine; malabsorption as a result of mucosal damage may be induced by drugs such as colchicine, neomycin and methotrexate. Glucocorticoids, diuretics and laxatives cause calcium depletion. Tetracyclines bind to calcium in the bones and lead to pigmentation of teeth. Cellulose phosphate and antacids (e.g. aluminium hydroxide) may render calcium unavailable. Cholestyramine may increase faecal loss of calcium. Persons on a vegan diet may have an insufficient calcium intake.

Table 5.9. Indices of iron status for adults aged 18–65 years

Plasma/serum	
Fe	
Men	0.7 – 1.8 mg/l; 12.5–32.2 μmol/l
Women	0.6 – 1.8 mg/l; 10.7–32.2 μmol/l
TIBC	44.8 – 80.6 μmol/l; 2.5–4.5 mg/l
% saturation	
Men	20% – 50%
Women	15% – 50%
Ferritin	4 – 103 μg/l (mean of men 40; women 30; SD 23)
Transferrin	2.0 – 3.5 g/l (0.7×TIBC[a])
Red blood cells	
Protoporphyrin	170 – 770 μg/l
Whole blood	
Haemoglobin	
Men	8.6 – 10.9 mmol/l
Women	7.4 – 9.6 mmol/l
Haematocrit	
Men	40 – 50 l/l
Women	36 – 45 l/l
MCV	84 – 104 fl
MCH	1750 – 2200 amol
MCHC	19.5 – 22.5 mmol/l

[a] TIBC, total iron-binding capacity.

Iron

The normal human body contains about 4 g of iron (mol.wt 55.85). It is present in several enzymes responsible for electron transport (cytochromes), for the activation of oxygen (oxidases and oxygenases) and for the transport of oxygen (haemoglobin in the bloodstream and myoglobin in the muscle cells). About 60%–70% of the body iron is found in haemoglobin and myoglobin. Approximately 20% is stored in a labile form as ferritin and haemosiderin in liver, spleen, bone marrow, and other tissues where iron can be used for the regeneration of haemoglobin in case of blood loss. The remaining 10% is firmly fixed in tissues.

The frequently used indicators for the assessment of the iron status are haemoglobin, haematocrit, serum iron, ferritin, and transferrin (Table 5.9). Levels indicative of iron deficiency are well described (WHO 1975). During the first stage of depletion the iron body store (excluding Hb) falls below 100 mg and serum ferritin levels drop below 12 μg/l. In the second stage transferrin decreases as a result of exhaustion of iron body store and/or increasing free protoporphyrin in red blood cells (subclinical iron deficiency). The third stage shows decreasing haemoglobin levels and is the stage of iron-deficiency anaemia.

Primary deficiency results from dietary insufficiency, i.e., a lack of iron, with or without sufficient vitamin C to facilitate the absorption of non-haem iron. Vegan or vegetarian diets, often in combination with parasitic and other infectious diseases, may induce an inadequate iron status. Secondary iron deficiency may originate from mucosal damage resulting from neomycin and colchicine use, and from substantial losses of iron from the body through bleeding, pregnancy, chronic dialysis and other routes. The absorption of iron from the gut can be affected by long-term use of antacids, tetracycline or cholestyramine. Phytic acid inhibits the absorption of vitamin C from the gut. The deficiency sign is iron-deficiency anaemia, with clinical symptoms in an advanced stage only.

Magnesium

The adult human body contains about 25g of magnesium (mol.wt 24.31); about 60% is present in the skeleton. Magnesium ions play an important role in a large variety of enzymatic reactions in which phosphate groups are involved. Furthermore, it functions in the metabolism of DNA and in the maintenance of macromolecular structures such as DNA and ribosomes. It is especially involved in the function of nervous, muscular, cardiac and vascular systems.

Plasma or serum levels (normal levels for adults 15–23 mg/l; 30% is bound to albumin) may remain constant in spite of a 20% reduction in body store. Almost all magnesium is intracellular. Red blood cell magnesium (50 ± 5 mg/l) is an insensitive indicator of magnesium status (probably as a result of blood cell formation in the skeleton). Magnesium in nails and hair does not show a good relation with its body store either. In cerebrospinal fluid levels of 24–34 mg/l have been found. The urinary Ca/Mg ratio should be <8. High dietary levels of calcium, phosphate, free fatty acids, dietary fibre and phytic acid inhibit absorption of magnesium. A high retention of Mg after parenteral loading indicates an inadequate supply (Caddell 1980). Kidneys restrict magnesium excretion to preserve the levels and, therefore, clinical signs of magnesium deficiency develop late (Shils 1964).

Primary deficiency may be induced by inappropriate slimming diets, which are low in magnesium. Secondary deficiency has been described as a result of disorders of the alimentary tract or of the kidneys, or by hormonal disturbances (Flink 1976). Magnesium deficiency is sometimes suspected in coronary heart diseases. Cellulose phosphate and antacids (such as aluminium hydroxide) may render magnesium unavailable. Increased excretion is induced by oral diuretics as well as by nephrotoxic agents. Acute deficiency signs include low plasma magnesium, general malaise, growth retardation, neuromuscular symptoms, heart rate disturbances, and stomach cramps.

Phosphorus

The adult human body contains about 700 g of phosphorus (mol.wt 30.97). Inorganic phosphates contribute structurally (mainly as hydroxyapatite) to bones, teeth and soft tissues and are involved in the maintenance of the acid–base balance within the cells and in the circulating plasma. Organic phosphates such as phospholipids are structural components of all cells and cofactors (e.g., in ATP) of various enzymatic reactions.

The principal method for the assessment of the phosphorus status is the analysis of phosphate (PO_4) in plasma where the levels decrease until middle age. Levels in adults are 0.8–1.5 mmol/l (25–47 mg/l). Urinary excretion depends on dietary intake. Normal levels are 500–600 mg phosphorus/24 h (1.5–1.8 g phosphate/24 h).

As phosphorus is abundant in nature, phosphorus deficiency as a result of dietary insufficiency is unknown. Secondary phosphorus deficiency may result from excessive intake of antacids. It has been suggested that the high phosphate/calcium ratio in the modern diet increases the risk of osteoporosis. As renal function decreases with age, a high phosphate intake may increase the risk of bone demineralization at older age. Deficiency signs include pain in muscles and bones, general malaise, haemolytic anaemia and kidney stones.

Potassium

Approximately 98% of total body potassium (mol.wt 39.10) (about 175 g for the adult) is located within the cells. Potassium is the principal intracellular cation with a concentration more than 30 times higher than in the extracellular fluid. Together with sodium and chloride, potassium is the main factor responsible for fluid osmolarity (osmotic pressure); these ions also influence ionic strength and, thus, the solubility of proteins and other constituents. Potassium and sodium are involved in muscle contractility and the conduction of impulses in nerves.

Potassium is analysed preferably in plasma: serum levels are somewhat higher as a result of blood cell release of potassium. In adults, levels of 3.5–4.5 mmol/l plasma are observed. Urinary excretion of potassium strongly depends on dietary intake. In adults on an average diet the excretion is 25–125 mmol/24 h.

Potassium deficiency may result from excessive loss as a result of infectious or nutritional diarrhoea, from diabetic acidosis, or from the use of alcohol and drugs like diuretics, glucocorticoids and laxatives. Increased loss due to mucosal damage may result from colchicine or neomycin use. Thiazide (a diuretic) impairs gastrointestinal transport. Hypokalaemia is also found in Cushing's disease (hypersecretion of adrenal steroids).

Trace Elements

Arsenic

The normal human body probably contains 8–20 mg arsenic (mol.wt 74.92) which is more or less equally distributed between various tissues. The highest levels are found in hair, nails and skin, probably resulting from its high affinity for binding to keratin. To date, no specific mode or site of action of arsenic has been found and it is still unclear whether arsenic is essential to human life (Nielsen 1980). In animals, it has been demonstrated that arsenic deficiency results in growth retardation, disturbed reproduction, leg abnormalities and elevated haematocrit and haemoglobin. The estimated daily intake of humans is about 10 μg/day. Observed levels in urine are below 50 μg/l.

Chromium

The total amount of chromium (mol.wt 52.00) in the normal human body is estimated at 1–10 mg. Before birth, this trace element is stored in the liver. It is needed for the maintenance of normal glucose tolerance, and trivalent chromium probably acts as a cofactor of insulin or is involved in the synthesis of the glucose tolerance factor. Lack of chromium results in reduced glucose tolerance and increased serum cholesterol and triglyceride levels, turbid cornea and plaques in the aorta, resembling the situation seen in diabetes. Peripheral neuropathy has also been described. However, a clear definition of chromium deficiency is lacking. Deficiency has been described in a few cases of parenteral feeding.

No reliable tests are available to define body chromium status. Large variations are found in biological material. Levels observed are for serum 0.08–45 μg/l, red blood cells 19–66 μg/l, whole blood 0.7–28 μg/l, hair 0.1–3.6 μg/g, and for urine less than 0.1 μg/24 h. There is a high risk of contamination interfering with analyses.

Cobalt

The normal human body contains about 1.5 mg of cobalt (mol.wt 58.93) of which only about 10% is present in vitamin B12. The only known function of cobalt is its role as an intrinsic part of this vitamin, the cobalt content being 4.34%. Man must meet his cobalt requirement through vitamin B12 intake. Cobalt deficiency and its causes seem to coincide with deficiency of vitamin B12.

Copper

The adult human body contains 50–120 mg copper (mol.wt 63.55), mainly located in the muscles and bones, with the highest concentrations in liver and hair. In plasma, copper is predominantly bound to its transport protein ceruloplasmin which functions as ferroxidase. Copper is also a cofactor of several other oxidases, for example superoxide dismutase and cytochrome oxidase (Prasad and Oberleas 1976).

The copper status is often investigated by analysis of its level in plasma or serum (see below) or red blood cells (about 0.7–1.1 mg/l) and of the plasma level of (the oxidase activity of) ceruloplasmin (90–510 mg/l) which contains 90%–95% of copper. However, it should be realized that the plasma copper level is highest in the morning and, furthermore, that a number of factors influence its level: age (infants 0.2–0.7 mg/l serum, children 6–12 years 0.8–1.9 mg/l), gender (men 0.7–1.4 mg/l, women 0.8 1.55 mg/l), hormones and pregnancy (1.2–3.0 mg/l). Copper in leucocytes would be a more reliable index (Bunker et al. 1984); levels are 4–37 pmol/10^6 leucocytes (men 12.5±6.7; women 11.7±6.7). In fat-free tissue 1.5–2.5 mg/kg is found. Other methods of indexing copper deficiency include the measurement of copper in the most proximal segments of hair, in urine (however, normal in Menkes' syndrome) and cuproenzyme activities (red blood cell-SOD and leucocytic cytochrome oxidase, and carboxypeptidase and 5′-nucleotidase) (Bunker et al. 1984, Solomons 1979).

An antagonism exists between copper and zinc. Especially in malabsorption, copper insufficiency is seen at a Zn/Cu intake ratio >10 or at a zinc intake above 50 mg/day. Furthermore, dosages of 1500 mg vitamin C reduce Cu and ceruloplasmin levels within the normal range. Copper deficiency is most likely to result from lack of intake, from a milk-based diet, (as milk is low in copper), or in the case of parenteral feeding for premature or low-birth-weight infants. A marginal copper status has been described for children with chronic malabsorption or chronic diarrhoea. Deficiency may develop in kidney dialysis. Deficiency may partly be induced or enhanced by excessive urinary excretion probably by chelation of copper when free amino acids are infused. Defective synthesis of ceruloplasmin is known as Wilson's disease and results in accumulation of copper in liver, brain and other tissues and a defective excretion of copper in bile. A disturbed copper absorption is present in Menkes' syndrome (kinky hair syndrome). Penicillamine induces copper deficiency by chelation. Deficiency signs are hypochromic anaemia with neutropenia, osteoporosis with vitamin C deficiency-like bone disorders, reduced skin pigmentation, or neurological symptoms (Wilson, Menkes).

Fluorine

The human body contains about 2–4 g of fluorine (mol.wt 19.00) of which most is found in bone, where it forms fluoroapatite from hydroxyapatite. This is important for hardening of tooth enamel and contributes to the bone mineral matrix. It may also provide "nucleation sites" for crystallization of the bone mineral and enhance bone density (WHO 1970).

Numerous studies have shown a negative correlation between the natural (or added) fluoride content of drinking water and the percentage of tooth decay (dental caries) in a population and a positive correlation between fluoride and bone density. Whether fluorine deficiency is a risk factor for osteoporosis is not yet clear. Methods for a reliable description of the fluorine status of an individual have not yet been developed. In adults, levels of 10–370 μg/l in serum and plasma, 450 μg/l in red blood cells and 0.2–1.9 mg/l in urine have been found.

Iodine

The iodine (mol.wt 126.90) content of the normal human body is about 10–20 mg. Iodine forms an integral part of the thyroid hormones (thyronines T_3 and T_4) which are essential to normal growth, development and function of the nervous system. The specific functions of the thyroid hormones are poorly understood, although it is clear that they do have effects on oxygen consumption and metabolic rate. Investigation of the iodine status is hampered by the fact that for an accurate analysis of urinary excretion of iodine at least three 24 h urine samples are needed. Average levels observed in healthy Dutch men (n= 380) and women (n=444) in the age range 18–92 years (Odink et al. 1988) were 1000 nmol/24 h (58.1 μmol/mol creatinine) and 797 nmol/24 h (62.1 μmol/mol creatinine), respectively.

Insufficient dietary intake is mainly a result of soils poor in iodine (spread worldwide). Secondary deficiency may result from a high intake of goitrogenic substances in the diet and from long-term use of drugs such as sulphonylureas, phenylbutazone, cobalt and lithium, which have been reported to impair the absorption and utilization of iodine, with a subsequent development of goitre.

Manganese

The adult human body contains about 10–20 mg of manganese (mol.wt 54.94). Its function is still poorly understood. It seems to be involved in several enzyme systems of carbohydrate and fat metabolism. It can activate a large number of metal–enzyme complexes. Other metals such as magnesium, however, may substitute for manganese in most instances. Manganese enzymes are present especially in the mitochondria.

Reliable methods for the assessment of manganese status are not yet known. There is a high risk of contamination. In experimental studies mitochondrial SOD (Mn enzyme not inhibited by cyanide, in contrast to Cu-Zn SOD) might be used. In adults the following levels are reported: 4–24 μg/l whole blood, 0.4–14 μg/l serum, 5–46 μg/l red blood cells, 1–50 μg/l urine, and 0.2–2.1 μg/g hair.

Manganese deficiency is unknown. An unproved manganese deficiency has been described in one man. Findings noted in that man were hypercholesterolaemia, weight loss, transient dermatitis, occasional nausea and vomiting, changes in hair and beard colour, and slow growth of beard and hair.

Molybdenum

The normal human body contains about 9–16 mg molybdenum (mol.wt 95.94). By analogy with animal tissues, the human body probably also contains some enzymes that use molybdenum for their activity (xanthine, aldehyde and sulphite oxidases). It can act as a copper antagonist. Molybdenum deficiency in man has not yet been described. The following levels in adults have been reported: 1–15 μg/l whole blood, 18 μg/l red blood cells (mean), 0.1–6 μg/l plasma/serum, 10–16 μg/l urine, and 0.06–0.20 μg/g hair. Contamination may be substantial.

Nickel

Nickel (mol.wt 58.70) is an essential nutrient for animals and probably also for man. The adult human body contains about 10 mg of nickel, with the largest proportions in skin and bone. It stimulates certain enzymes (e.g., arginase and deoxyribonuclease) and is probably involved in fat metabolism and the maintenance of the structure of nucleic acids and membranes, especially of mitochondria. Furthermore, nickel is claimed to be involved in the absorption of iron (as its ferric ion).

Nickel deficiency in man is unknown. Low serum levels are found in patients with liver cirrhosis and chronic uraemia. Normal levels are 1–7 μg/l serum. A nickel-binding protein (nickeloplasmin) is found in plasma (Nielsen 1980).

Selenium

The adult human body contains about 20 mg of selenium (mol.wt 78.96). Based on animal studies, it is assumed that selenium is essential to man as well. The best known function of selenium is its contribution to glutathione peroxidase (GSH-Px), which protects vital components of the cell against oxidative damage. Furthermore, through binding and formation of complexes with protein, selenium is said to be protective against poisoning by heavy metals such as lead, cadmium and mercury. Selenium plays a poorly understood role in growth, preserving muscle function, integrity of the liver, and fertility. Selenium is closely, but not completely, associated with vitamin E. The vitamin and the trace element appear to be synergistic in correcting deficiency states, probably because both function as antioxidants.

Levels in blood cells and plasma are strongly influenced by dietary intake which depend on the selenium content of the soil. Whole blood and blood cell levels change more slowly than plasma levels. Liver selenium changes rapidly after a change of intake. In general, the following levels are reported: 57–340 μg/l whole blood, 78–320 μg/l plasma and serum, 71–340 μg/l red blood cells, 5–100 μg/l urine, and 0.6–2.6 μg/g hair. In areas with a low level of selenium in soil, such as Canada and New Zealand, and formerly in Finland (at present fortified fertilizers are used), whole blood levels are about 47–60 μg/l. Selenium contents in urine, hair and nails vary as a result of exogenous factors, such as selenium-containing shampoo.

The GSH-Px activity in blood cells correlates with blood selenium up to 100 μg/l. The level is best measured in platelets. After selenium supplementation, platelet GSH-Px is a good indicator of the conversion in tissue of selenium into a biologically active form. Plasma GSH-Px originates from the liver (Chen Shang et al. 1982; Robberecht and Deelstra 1984). Plasma contains as yet unidentified binding proteins.

Disease states attributable to selenium deficiency have not yet been described in man, even though large population groups are living in areas where selenium deficiency has resulted in severe disease in livestock. An exception might be "Keshan disease", observed in several areas of China. Keshan disease is prevented by supplementation, but not cured. Low blood levels have been observed in total parenteral nutrition and in patients with gastro-intestinal cancer, during pregnancy, and PEM. High levels are found in patients with reticulo-endothelial neoplasia. Kwashiorkor probably points to a low selenium intake (Van Rij et al. 1979; Mathias and Jackson 1982). Deficiency signs are muscle pain and weakness, white nail matrix, and decreased GSH-Px activity.

Silicon

The human body probably contains gramme quantities of silicon (mol.wt 28.09) with the highest concentration in the skin. High concentrations are also found in the arterial wall, especially in the intima. From animal studies it is clear that silicon is necessary for normal growth and development and optimum bone mineralization. It is involved in the formation of glycosaminoglycans and collagen of the matrix and is a basic substance of bone, articular cartilage and connective tissue. It appears to be the major anion in osteogenic cells. Deficiency of silicon in humans is not known.

Tin

Tin (mol.wt 118.69) in the normal human body amounts to about 14 mg; the highest concentrations are found in teeth and skeleton. It is probably important for the structure of proteins and other macromolecules. Deficiency of tin in humans is unknown. In animals, deficiency of tin results in growth retardation.

Vanadium

The normal human body contains about 10 mg of vanadium (mol.wt 50.94), mainly in the teeth. It seems to have a positive effect on bone mineralization, synergistic with silicon. It may have a specific physiological role in the transport of sodium and potassium across the cell membrane. Furthermore, it has been suggested that vanadium is involved in fat metabolism and cholesterol synthesis. In man, signs of vanadium deficiency are not known. It has been suggested that a shortage of vanadium is involved in kwashiorkor and in oedema of children with PEM. Reported levels in serum are 350–480 μg/l.

Zinc

The average content of zinc (mol.wt 65.38) in the normal human body is about 1.6–2.3 g; 65% is located in the muscles, 25% in the skeleton, 2% in blood and the remainder in liver, skin, kidneys and pancreas. In blood, 80% of zinc is within the red blood cells. Zinc is a component of over 200 enzyme systems including carbonic anhydrase, alkaline phosphatases, dehydrogenases and carboxypeptidases. It has a function in taste, in the immune system and in the brain (Prasad and Oberleas 1976, Prasad 1979).

Methods are not yet available for a reliable assessment of zinc status. However, a series of indices are used such as the analysis of zinc in plasma, serum, red blood cells, leucocytes (Bunker et al. 1984), hair, urine and saliva, and functional tests measuring the zinc metalloenzyme activities and taste acuity. Zinc in plasma is mainly associated with its carrier albumin. Hypoalbuminaemic conditions (hepatic cirrhosis, malnutrition) show reduced plasma zinc levels. Reduced levels are also found by the action of bacterial endotoxin and by various forms of steroids (e.g., oral contraceptives). Reported zinc levels are: 0.7–1.5 mg/l plasma (1.12±0.12 mg/l; with circadian and postprandial fluctuations), 1.3–16.3 mg/l red blood cells, 193±18 μg/g hair. Deficiency of zinc is present at fasting plasma levels <0.7 mg/l (the serum level is 16% higher than the plasma level as a result of release by platelets). A deficiency in pregnancy exists at <0.5 mg/l plasma. The zinc level of hair is susceptible to a number of factors: a supposed suboptimal status at <100 μg/g hair, and deficiency at <70 μg/g (Solomons 1979). Possible contamination renders 24 h urine unsuitable as a reliable indicator of the zinc status.

Carboxypeptidase and 5′-nucleotidase activities seem reliable indicators of zinc nutritional status, although their final value awaits proof. Alkaline phosphatase activity is not promising because it varies widely and is influenced by many factors. Biochemical markers that still have to be evaluated are the levels of zinc in red blood cells and leucocytes, in saliva, hair, bone and muscles (zinc levels are threefold higher in red muscle than in white muscle), and probably functional taste tests.

Definite conclusions are based upon the effects of oral zinc supplementation on the forementioned indicators (Halsted et al. 1974).

Deficiency has been described as a result of malabsorption, liver and kidney disorders, excessive losses (e.g., uraemia, dialysis, burns), inhibition of absorption by dietary fibre and phytic acid (for the latter significant at a phytic acid/zinc ratio higher than 12), long-term drug use (diuretics such as thiazides and frusemide, chelating agents such as penicillamine, cancer chemotherapeutic agents such as cisplatinum, laxatives) and alcohol abuse. Severe deficiency results in hypogonadism, hepatosplenomegalia, open epiphysis, reduced taste and smell, nail and skin abnormalities, iron deficiency anaemia, and night-blindness.

Discussion

Most biochemical markers of the micronutrient status require the analysis of levels of micronutrients or of some of their active metabolites or catabolites in only a few biological matrices (mostly whole blood, plasma, serum or urine). Only a limited number of markers seem valuable for the evaluation of the (long-term) status at the cellular level or at the level of the entire organism (e.g., enzyme activity and/or stimulation tests) when special circumstances are taken into account (e.g., disease, drug use).

However, one should distinguish between the evaluation of the micronutrient status at the individual level and at the level of a population. In surveys, trends, risk evaluation, correlations (intake versus status) and group comparisons are often investigated. In these situations, several biochemical markers, such as urinary excretion levels, are valuable, whereas they prove less reliable at the individual level. Furthermore, for surveys the methods used for the evaluation of the nutritional status are more practical than for individual (e.g., patient) studies. Nevertheless, both types of study lack reliable and/or practical biochemical markers for several vitamins (vitamins A, E and K, niacin, pantothenic acid and biotin) and a number of macroelements (e.g., calcium and magnesium) and trace elements such as chromium, manganese and zinc.

Although functional indices of the nutritional status (as thoroughly reviewed by Solomons and Allen 1983) can be of great value especially at the individual level, they are usually not very practical as they generally require repeated sampling and analysis. With respect to the matrix for an analysis of micronutrients, several matrices other than blood and urine should be evaluated further. These include hair and nails with respect to trace elements, and fat biopsy with respect to the fat-soluble vitamins such as carotenoids and vitamin E, the latter especially in relation to polyunsaturated fatty acids.

However, it is to be expected that functional indices of the micronutrient status at the level of tissues, organs and/or the entire organism will never fully replace the direct, so-called static biochemical markers. In general, one will need information of the micronutrient itself, i.e. the quantitative level of a micronutrient in the matrix (Van der Beek et al. 1988). The ultimate functional indices will describe the micronutrient-related process at a qualitative level. The highest potential of these indices can be the evaluation of the specificity and sensitivity of the biochemical markers of the nutritional status, i.e. to validate cut-off levels.

References

Bamji MS (1981) Laboratory tests for the assessment of vitamin nutritional status. In: Briggs MH (ed) Vitamins in human biology and medicine. CRC Press, Boca Raton, Florida, pp 1–27

Basu TK (1988) Drug-nutrient interactions. Croom Helm, London

Beaton GH (1986) Toward harmonization of dietary, biochemical, and clinical assessments: The meanings of nutritional status and requirements. Nutr Rev 44:349–358

Boeschoten EW, Schrijver J, Krediet RT, Schreurs WHP, Arisz L (1988) Deficiencies of vitamins in CAPD patients: the effect of supplementation. Nephrol Dial Transplant 2:187–193

Bouillon RA, De Laey P (1983) Vitamin D status in man. Institut National des Radioelements, Fleurus, Belgium

Bunker VW, Hink LJ, Lawson MS, Clayton BE (1984) Assessment of zinc and copper status of healthy elderly people using metabolic balance studies and measurement of leucocyte concentrations. Am J Clin Nutr 40:1096–1102

Caddell J (1980) Clinical signs of magnesium deficiency. Nutr Rev 38:100–102

Chen Sang Y, Collipp PJ, Boasi LH, Isechnschmid, Verolla RJ, San Roman GA, Yeh JK, (1982) Fluorimetry of selenium in human hair, urine and blood. Ann Nutr Metab 26:V+—Ü–+&?

Christakis G (ed) (1972) Nutritional assessment in health programs. American Public Health Association, Washington DC

Das KC, Herbert V (1978) The lymphocyte as a marker of past nutritional status: persistence of abnormal deoxyuridine (dU) suppression test and chromosomes in patients with past deficiency of folic acid and vitamin B12. Br J Haematol 38:219–233

Das KC, Herbert V, Coleman N (1978) Unmasking covert folic acid deficiency in iron-deficient subjects with neutrophil hypersegmentation: dU suppression tests on lymphocytes and bone marrow. Br J Haematol 39:357–375

Debry G (ed) (1984) Nutrition, food and drug interactions in man. Wld Rev Nutr Diet, vol 43

Devgun MS (1981) Vitamin D nutrition in relation to season and occupation. Am J Clin Nutr 34:1501–1504

Ellegaard J, Esmann V (1970) A sensitive test for folic acid deficiency. Lancet I:308

Fidanza F (ed) (1986) Nutritional status assessment methodology for individuals and population groups. Group of European Nutritionists (GEN), Workshop Colombella, Perugia, Italy

Flink EB (1976) Magnesium deficiency and magnesium toxicity in man. In: Prasad AS, Oberleas D (eds) Trace elements in human health and disease, vol II, Essential and toxic elements. Academic Press, New York, pp 1–21

Friedrich W (1988) Vitamins. Walter de Gruyter, Berlin

Golden MHN (1983) Trace elements in human nutrition. Hum Nutr Clin Nutr 36C:185–202

Guillaumont M, LeClercq M, Gosselet H, Makala K, Vignal B (1988) HPLC determination of serum vitamin K1 by fluorometric detection after post-column electrochemical reduction. J Micronutr Anal 4:285–294

Halsted JA, Smith Jr JC, Irwin MI (1974) A conspectus of research on zinc requirements of men. J Nutr 104:345–378

Hambridge KM, Mauer AM (1978) Trace elements. In: NRC/FNB Laboratory indices of nutritional status in pregnancy. National Academy of Sciences, Washington DC, pp 157–194

Hathcock JN, Coon J (eds) (1978) Nutrition and drug interrelations. Academic Press, New York

Huybers WAR, Schrijver J, Speek AJ, Deelstra BA, Okken A (1986) Persistent low plasma vitamin E levels in premature infants surviving respiratory distress syndrome. Eur J Pediatr 145:170–171

Kübler W (1988) Häufigkeit und physiologische Bedeutung von Vitaminmangelzustanden. In: Wolfram G, Schlierf G (eds) Ernährung und Gesundheit. Beiträge der Ernährungsepidemiologie in Europa. Wissenschaftliche Verlagsgesellschaft, Stuttgart, pp 161–176

Lambert WE, DeLeenheer AP, Baert EJ (1986) Wet-chemical postcolumn reaction and fluorescence detection analysis of the reference interval of endogenous serum vitamin $K_{1(20)}$. Anal Biochem 158:257–261

Lindeman RD (1984) Assessment of trace element depletion. In: Wright RA, Heymsfield S (eds) Nutritional assessment. Blackwell Scientific Publications, Oxford, pp 239–261

Mathias PM, Jackson AA (1982) Selenium deficiency in kwashiorkor. Lancet I:1312–1313

Mobarham S, Russell RM, Underwood BA, Wallingford J, Mathieson RD, Al-Midiani (1981) Evaluation of the relative dose response test for vitamin A nutriture in cirrhotics. Am J Clin Nutr 34:2264–2270

Natadisastra G, Wittpenn JR, Muhilal, West Jr KP, Mele L, Sommer A (1988) Impression cytology: a practical index of vitamin A status. Am J Clin Nutr 48:695–701

Nielsen FH (1980) Evidence of the essentiality of arsenic, nickel, and vanadium and their possible nutritional significance. Adv Nutr Res 3:157–172

Odink J, Bogaards JJP, Sandman H, Egger RJ, Arkesteyn, De Jong P (1988) Excretion of iodide in 24 h urine as determined by ion-pair reversed-phase liquid chromatography with electrochemical detection. J Chromatogr 431:309–316

Olsen JA (1982) New approaches to methods for the assessment of nutritional status of the individual. Am J Clin Nutr 35:1166–1168

Ovesen L (1979) Drugs and vitamin deficiency. Drugs 18:278–298

Prasad AS (1979) Zinc in human nutrition. CRC Press, Boca Raton, Florida

Prasad AS (ed) (1982) Clinical and nutritional aspects of trace elements. Alan R Liss, New York
Prasad AS, Oberleas D (eds) (1976) Trace elements in human health and disease, vol 1: Zinc and copper, vol 2: Essential and toxic elements. Academic Press, New York
Rennert OM (ed) (1984) Metabolism of trace elements in man. Vol I: Developmental aspects, vol II: Genetic implications. CRC Press, Boca Raton, Florida
Robberecht HJ, Deelstra HA (1984) Selenium in human urine: concentration levels and medical implications. Clin Chim Acta 136:107–120
Roe DA (1984) Nutrient and drug interactions. Nutr Rev 42:141–154
Sauberlich HE (1978) Vitamin indices. In: NRC/FNB Laboratory indices of nutritional status in pregnancy. National Academy of Sciences, Washington DC, pp 109–156
Sauberlich HE (1983) Current laboratory tests for assessing nutritional status. Surv Synth Path Res 2:120–133
Sauberlich HE (1984a) Implications of nutritional status on human biochemistry, physiology, and health. Clin Biochem 17:132–142
Sauberlich HE (1984b) Newer laboratory methods for assessing nutriture of selected B-complex vitamins. Ann Rev Nutr 4:377–407
Sauberlich HE (1986) Methods for the assessment of nutritional status. In: Chen LH (ed) Nutritional aspects of aging, vol I. CRC Press, Boca Raton, Florida, pp 132–157
Sauberlich HE, Dowdy RP, Skala JH (1974) Laboratory tests for the assessment of nutritional status. CRC Press, Cleveland, Ohio
Schrijver J, Speek AJ, Schreurs WHP (1981) Semi-automated fluorometric determination of pyridoxal-5′-phosphate (PLP) in whole blood by high-performance liquid chromatography (HPLC). Int J Vit Nutr Res 51:216–222
Schrijver J, Speek AJ, Klosse JA, Van Rijn HJM, Schreurs WHP (1982) A reliable semiautomated method for the determination of total thiamine in whole blood by high-performance liquid chromatography. Ann Clin Biochem 19:52–56
Schrijver J, Van Schaik F, Van der Greef J, Bos KD, Schreurs WHP (1984) HPLC-analysis of xanthurenic acid and its 8-methyl ether. In: Schlossberger HG, Kochen W, Linzen B, Steinhart H (eds) Progress in tryptophan and serotonin research. Walter de Gruyter, Berlin, pp 71–75
Schrijver J, Van Veelen WC, Schreurs WHP (1985) Biochemical evaluation of the vitamin and iron status of an apparently healthy Dutch free-living elderly population. Comparison with younger adults. Int J Vit Nutr Res 55:337–349
Selhub J, Rosenberg IH (1984) Assessment of vitamin depletion. In: Wright RA, Heymsfield S (eds) Nutritional assessment. Blackwell Scientific Publications, Oxford, pp 209–238
Shils ME (1964) Experimental magnesium depletion I. Clinical observations and blood chemistry alterations. Am J Clin Nutr 15:133–143
Solomons NW (1979) On the assessment of zinc and copper nutriture in man. Am J Clin Nutr 32:856–871
Solomons NW, Allen LH (1983) The functional assessment of nutritional status: principles, practice and potential. Nutr Rev 41:33–50
Speek AJ, Van Schaik F, Schrijver J, Schreurs WHP (1982) Determination of the B2 vitamer flavin-adenine dinucleotide in whole blood by high-performance liquid chromatography with fluorimetric detection. J Chromatogr 228:311–316
Speek AJ, Schrijver J, Schreurs WHP (1984) Fluorimetric determination of total vitamin C in whole blood by high-performance liquid chromatography with pre-column derivatization. J Chromatogr 305:53–60
Speek AJ, Van Agtmaal EJ, Saowakontha S, Schreurs WHP, Van Haeringen NJ (1986) Fluorimetric determination of retinol in human tear fluid using high-performance liquid chromatography. Curr Eye Res 5:841–845
Underwood EJ (1977) Trace elements in human and animal nutrition, 4th edn. Academic Press, New York
Van der Beek EJ, Van Dokkum W, Schrijver J et al. (1988) Thiamin, riboflavin, and vitamins B6 and C: impact of combined restricted intake on functional performance in man. Am J Clin Nutr 48:1451–1462
Van Rij AM, Thomson CD, McKenzie JM, Robinson MF (1979) Selenium deficiency in total parenteral nutrition. Am J Clin Nutr 32:2076–2085
WHO (1970) Fluorides and human health. Monograph series no. 59. WHO, Geneva
WHO (1973) Trace elements in human nutrition. Technical report series no. 532. WHO, Geneva
WHO (1975) Control of nutritional anaemia with special reference to iron deficiency. Technical report series no. 580. WHO, Geneva

Whyte MP (1979) Vitamin D bioavailability: serum 25-hydroxyvitamin D levels in man after oral, subcutaneous, intramuscular and intravenous vitamin D administration. J Clin Endocrinol Metab 48:906–911
Winick M (ed) (1983) Nutrition and drugs. Curr Conc Nutr, vol 12
Wright RA, Heymsfield S (eds) (1984) Nutritional assessment. Blackwell Scientific Publications, Oxford
Ziegler R (1988) Osteoporosis, calcium und vitamin D. In: Wolfram G, Schlierf G (eds) Ernährung und Gesundheit. Beiträge der Ernährungsepidemiologie in Europa. Wissenschaftliche Verlagsgesellschaft, Stuttgart, pp 156–160

Chapter 6

The Functional Significance of Marginal Micronutrient Deficiency

R. Buzina and K. Subotičanec

Nutrition surveys conducted during the past 10 years have shown that broad segments of populations even in the more developed, industrial countries are not obtaining their recommended dietary allowances (RDA) of many micronutrients from the foods they are consuming. The available data show that such inadequate dietary intake may affect the vitamin and mineral nutritional status in the general population as well as in some selected, more vulnerable groups. In this respect data from Switzerland, Federal Republic of Germany, France and Hungary are of particular interest: studies on the nutritional status of large population segments have been carried out in those countries (Aebi et al. 1984, Ernährungsbericht 1984, ESVITAF 1986, Biro G. personal communication, 1987). Tables 6.1 to 6.4 summarize data on the prevalence of inadequate vitamin nutritional status based primarily on biochemical criteria.

The data presented indicate that subclinical vitamin deficiencies do occur in almost all the populations examined. Folate and vitamin A deficiencies were most common, followed by thiamin, pyridoxine, riboflavin and vitamin C. In Sweden and Norway, the dietary intake of vitamin D seems also to be often inadequate and between 36% and 76% of children aged 2–4 years were found to have a dietary vitamin D intake below 60% of the Swedish RDA (Hagman et al. 1986). Vitamin D deficiency was also reported in elderly subjects in Oslo so that during the winter about 40% of the free-living elderly women and between 77% and 83% of subjects of both sexes living in old people's homes had vitamin D plasma values below 20 μg/ml (SW Sem, personal communication, 1988).

With regard to mineral nutritional status, iron and calcium deficiencies seem to be quite common. In the Swiss study 15% of preschool children consumed less than 70% of the RDA for calcium and 36% of them consumed less than 70% of RDA for iron. For other population groups, inadequate iron intake, based on plasma iron determination was found particularly in military conscripts and affected about 11.5% of the examined subjects. Inadequate iron nutrition status was also found in French women of child-bearing age in the ESVITAF study. Only 11% of the examined women had the required iron intake. Iron

Table 6.1. Vitamin status in selected Swiss populations (from Zweiter Schweizerischer Ernährungsbericht 1984)

Age groups (yr)	Subjects with inadequate biochemical vitamin values (%)				
	B1	B2	B6	C	A
5–6	0.7				
7–10	2.0–3.1	0.0–2.0	11.8–12.1	0.0–2.3	7.0–12.0
11–14	8.3–11.8	2.4–3.9	18.8–21.2	1.0–5.5	
15–19	12.7–18.9	4.7–5.3	23.1–24.0	1.0–3.7	
⩾20	12.0–15.0	0–15	9.0–15.0	2.3–26.0	3.3–11.7

Table 6.2. Vitamin status in selected German populations (from Ernährungsbericht 1984. Deutsch. Gesell. Ernährung 1984)

Population	Subjects with inadequate biochemical values (%)						
	A	B1	B2	B6	C	Folic Acid	E
18–24 yr	3.9–25	13.6–20	5.8–22	7–10	2.2	10–14	0–9
65 yr	4.8–9.4	14.8–15	5.4–5.9	5.5–10.8	8.7–9.4	9.4–19.2	0.0–0.5
Pregnancy	8	14	13	14	4		

Table 6.3. Vitamin status in selected French populations aged 18–44 years (from ESVITAF 1986)

	Subjects with inadequate biochemical values (%)							
	Thiamine		Riboflavin		Folates		Vitamin A	
	Deficient	Borderline	Deficient	Borderline	Deficient	Borderline	Deficient	Borderline
Men	4.3	19.4	0.0	0.0	3.0	22.0	0.0	3.0
Women	10.3	28.7	0.0	0.0	2.0	21.0	0.0	30.0

Table 6.4. Vitamin malnutrition in selected Hungarian populations (from G. Biro FENS Congress Warszawa 1987)

	Subjects with inadequate biochemical values (%)					
	Male (yr)			Female (yr)		
	18–34	35–60	61	18–34	35–60	61
Vitamin A	31.0	39.1	46.2	62.1	52.8	46.9
Thiamin	5.0	2.4	3.4	9.8	5.8	3.0
Folates	12.0	11.4	9.1	11.5	8.6	11.8
Ascorbic Acid	32.6	35.4	27.0	29.7	21.4	32.2

stores were found to be zero in 13% and incomplete in the 75% of the women of child-bearing age.

Inadequate intakes of iron, calcium and iodine were also reported from the Federal Republic of Germany. With regard to iron, inadequate nutritonal status, defined as serum iron below 60 mcg/dl, was found in 13.5% of military conscripts, in 9.3% of young adults between 18 and 24 years and 6.9% of elderly subjects.

Infants and children, as well as pregnant and lactating women, are among population groups which are more sensitive to, and may be more affected by nutritional inadequacies. In the Swiss study it was reported that in a sample of pregnant women, about 78% had inadequate vitamin A, thiamin, pyridoxine, biotin and folic acid intakes. In the German study, on the basis of biochemical examinations, inadequate intakes of vitamin A, thiamin, riboflavin and pyridoxine were quite common in the third trimester of pregnancy. Van den Berg et al. (1978) have also reported a relatively high prevalence of marginal vitamin deficiencies especially pyridoxine, vitamin B12 and folic acid in Dutch parturient women considered to be healthy. Similar data have been reported by Jacob et al. (1976), Bailey et al. (1980) and Bowering et al. (1980). Schuster et al. (1981) have reported that pyridoxine deficiency in pregnant, low-income adolescent and adult women may affect the condition of the infant at birth. In connection with this, it should be mentioned that the previously held concept that the fetus is an active and efficient parasite extracting all essential nutrients at the expense of maternal stores has been replaced by the realization that the developing fetus can be influenced by a number of environmental factors including nutrition (Rush et al. 1980; Mora et al. 1979).

Another population group that may be particularly affected is the elderly. Studies of elderly people in Italy (Fidanza et al. 1984), Sweden (Steen et al. 1977), United Kingdom (Department of Health and Social Security 1979) as well as in the US (Kerr et al. 1982, Bidlack et al. 1986, McGandy et al. 1986) and Australia (Flint et al. 1979) have shown inadequate intakes, particularly of thiamin, riboflavin, pyridoxine and folic acid in most, and inadequate intakes of vitamin A in some of the examined populations. Studies on biochemical assessment of nutritional status in the elderly have also shown that chronic subclinical vitamin deficiencies of thiamin, riboflavin, pyridoxine, vitamin C and vitamin D are quite common in the elderly (Vir and Love 1978, Fisher et al. 1978, Contaldo et al. 1978, Suter and Russel 1987, Bouillon et al. 1987, Omdahl et al. 1982).

Aetiology of Micronutrient Deficiency

Adequate and balanced nutrition depends on the proper and varied selection of foods, as there is practically no single food (except breast milk for babies) containing all essential nutrients in adequate amounts. The major factor determining the choice of foods is dietary habits but in many population groups income and the price of foods may constitute a serious limiting factor in food selection. Foods containing higher amounts of essential nutrients such as milk and milk products, legumes, fresh fruits and vegetables, are usually more expensive than the energy-yielding cereals, rice, potatoes and fats, and therefore their consumption may not be adequate in low-income families. But, though low-income families may often not meet dietary recommendations, they are by no means the only ones who may be affected by inadequate micronutrient intake. Other factors which may affect the balance between the energy and the specific nutrient content of the diet are changes in dietary habits, particularly the increased consumption of refined, industrially processed foods, as well as the increased production and consumption of formulated, so-called convenience foods which are prepared to resemble standard recipes but are often of lower nutrient density. Furthermore, considering that the tendency toward reduction of physical activity in contemporary society has resulted in the reduction of total energy intake, the foods with lower micronutrient density per unit of energy may no longer meet the RDA even in population groups not considered to be poor. According to a US Department of Agriculture Report (1980), the percentage of diets providing less than 100% of the RDA for vitamins A, B1, B2, C and iron has more than doubled for most of those nutrients during the 22-year period in which these studies were conducted. The reason for this is the increased consumption of snacks and "fast food" or "convenience food". It was shown also that a typical fast-food lunch, which may supply about 14% of the recommended daily energy intake for a young adolescent, would supply only between 2% and 17% of the daily requirements for vitamin C, vitamin A, folic acid, niacin and pyridoxine (Birkbeck 1985).

Another population group which may also suffer micronutrient deficiencies independently of economic status are people consuming low energy diets to reduce or prevent overweight and obesity. A low energy diet may often not provide adequate amounts of essential micronutrients from natural sources (Mareschi et al. 1984). Such a situation could be particularly critical in overweight and obese pregnant women trying to reduce the normal weight gain in pregnancy.

People with special dietary habits may also be at risk of developing qualitative nutritonal deficiencies. So, for example, questions are raised whether vegetarianism may affect the micronutrient intake. While micronutrient intake is not a problem for persons on lacto-ovovegetarian diets, strict vegetarians (vegans) who exclude all foods of animal origin, including milk and eggs, may have inadequate intakes of vitamins B12, vitamin D, riboflavin and zinc. This problem could be even worse in people adhering to the extreme macrobiotic diet.

Numerous drugs have also been shown to cause micronutrient malabsorption and may affect their bioavailability and/or metabolism. Among these are mineral oils, antacids, some antibiotics and antitumour agents. Some drugs are also vitamin antagonists. Antivitamin properties are also recognized as side effects of isoniazid, hydralazine, cycloserine, laevodopa and penicillamine. It has been suggested that oral contraceptive steroids may affect riboflavin and pyridoxine status in women on marginal dietary vitamin intakes.

The consumption of alcoholic beverages may also contribute to the dilution of nutrient density in relation to total energy intake, and may interfere with gastro-intestinal absorption of nutrients. Vitamins commonly deficient among those ingesting high quantities of alcoholic beverages include folate, thiamin, ß-carotene and pyridoxine, but all other micronutrient deficiencies can be expected to occur in those who at the same time ingest restricted varieties and amounts of food.

Furthermore, data are available indicating that cigarette smoking interferes particularly with vitamin C status. Several studies have shown that smokers have lower plasma and leucocyte levels of vitamin C and can be considered to represent a risk group. To reach the ascorbate status of non-smokers, smokers would have to ingest daily 30–40 mg (up to 140% of the present RDA) more than non-smokers (Keith and Mossholder 1986).

Functional Significance of Marginal Micronutrient Deficiencies

In view of the relatively high prevalence of marginal, biochemically diagnosed micronutrient deficiencies, the question has often been raised about their effects on functional and health status in humans. Though it is well recognized that severe deficiency of specific micronutrients will impair human functions such as physical work capacity, immunocompetence, behaviour and cognition, more recent data also indicate that milder micronutrient deficiencies and, in particular, deficiencies that are reflected by biochemical findings that are not associated with obvious signs and symptoms may also affect human functions.

Van der Beek (1986), Van der Beek et al. (1987) and Spurr (1987) have recently reviewed the effects of restricted micronutrient intake leading to marginal micronutrient deficiency on physical work capacity (PWC). They indicated that of the minerals, iron showed the clearest relationship and concluded that there was direct evidence for the deleterious effect of moderate iron deficiency on PWC.

They concluded that there is some evidence for a relationship between subclinical vitamin deficiencies and PWC particularly in regard to thiamin, riboflavin and vitamin C. The existing data also show that effects on functional performance depend on relative restrictions of vitamin intake and duration of the restriction period. Studies in which the vitamin intake was furthest below RDA (20%–40%) have provided the strongest evidence for a deleterious effect on PWC.

In our recent studies we have also shown that biochemical vitamin C deficiency is associated with reduced aerobic capacity, one of the criteria of physical fitness, and that supplementation with physiological doses of ascorbic acid results in improved aerobic capacity. Since the populations studied also had a higher prevalence of biochemically defined riboflavin and pyridoxine deficiencies we examined the relationship between those two vitamins and aerobic capacity. The results show a positive and significant correlation between aerobic capacity and erythrocyte pyridoxine and a positive, but statistically not significant, correlation with erythrocyte riboflavin content. From the shape of the curves it can be concluded that the restricted intake of vitamin C, pyridoxine and riboflavin leading to marginal deficiencies may affect functional performance but that the intake of vitamins in amounts greater than the physiological requirements is not likely to improve physical performance.

The effect of marginal malnutrition on immune functions has been studied by many authors, and extensively reviewed by McMurray et al. (1981), Beisel (1982) and Chandra and Newberne (1977). It is suggested that malnutrition affects primarily the cell-mediated immunity whereas the humoral immune system appears to be better preserved in marginal and moderate malnutrition. Among vitamins, it was shown that vitamin A deficiency impairs complement functions and lymphocyte proliferation whereas folate deficiency results in decreased lymphocyte stimulation response to mitogens. Pyridoxine deficiency may impair cellular immunity as well but may also result in diminished antibody response to tetanus toxoid and typhoid vaccine (Hodges et al. 1962). Vitamin C seems to have many functions regarding the immune system. It enhances bacterial phagocytosis, and the addition of vitamin C to the cultures of neutrophiles or macrophages increases their mobility and chemotaxic activity. Vitamin C appears to be required by the thymus for the maintenance of reticular cells and may be associated with the synthesis of interferon.

With regard to minerals, zinc deficiency has a profound impact on cell-mediated immunity and to a lesser extent on other immune responses. Even marginal zinc deficiency induced in human volunteers results in impaired T-cell function and decreases serum thymic hormone activity (Chandra and Singh 1987). Iron deficiency also impairs cell-mediated immune responses, for example, delayed cutaneous hypersensitivity, T-cell number, lymphocyte proliferative responses to mitogens and antigens, and lymphokine production. Secretory IgA antibody response may also be deficient.

In a study recently conducted in institutionalized elderly (Subotičanec et al. 1989), we have also found a positive and significant correlation between delayed cutaneous hypersensitivity and biochemical parameters of iron, zinc, vitamin C and thiamin nutrition status whereas the association between the parameters of riboflavin and pyridoxal status, although positive, did not reach the level of statistical significance. The calculation of multiple correlation after inclusion of all biochemical parameters into the stepwise regression analysis, has shown that the coefficient of multiple regression between the biochemical parameters of nutritional status examined and delayed cutaneous hypersensitivity was $R=0.549$ ($P<0.01$) which indicates that about 36% of the variability in the cellular immunity would be affected by the existing vitamin and mineral nutrition status.

Since the immune system is an important host defence mechanism not only against infectious disease but also against a variety of other disorders such as cancer and autoimmune diseases, the impact of micronutrient deficiencies, particularly those that are commonly seen in selected population groups, requires further investigation.

Severe vitamin deficiency results in dramatic disturbances of behaviour, cognitive functions, emotional state and personality but subjects with subclinical vitamin deficiency also showed behavioural changes. The most common was an increased tendency to depression, irritability, lassitude and impairment of short term memory (Chome et al. 1984). Similarly Goodwin et al. (1983) have shown an association between poor performance in cognitive tests and low intake and serum level of riboflavin, folate, vitamin B12 and vitamin C in otherwise healthy elderly populations. Recent studies also suggest that nutrition may have specific effects on the synthesis of brain transmitters that are involved in a variety of behaviour such as sleep, mood, and depression. Several dietary components are neurotransmitter precursors, and pyridoxine, vitamin C and thiamin are associated with the enzymes critical in neurotransmitter synthesis.

With regard to the effects of mineral deficiencies on behaviour and cognition, data by Pollitt et al. (1986) are of particular interest. These indicate that infants with iron deficiency with and without anaemia tend to score lower on tests of mental development, and that iron repletion therapy results in an improvement in scores on a mental development scale. These results are in agreement with the general proposition that there are positive associations between somatic iron levels of infants and preschool children and their performance of cognitive function. Sandstead (1986) has recently reviewed the influence of trace elements on brain function and has concluded that iron, iodine, cobalt, copper, manganese, and zinc may have an important influence on brain development and function.

Results of a study we have recently conducted in school children aged 9–10 years in a rural area with high prevalence of biochemical deficiencies of vitamin C, riboflavin and pyridoxine have shown a small but statistically significant association between the nutritional status and cognitive functions, particularly perceptive speed. The results of stepwise multiple correlation show that about 16% of the variability of the cognitive test scores could be explained by the nutritional status of the children, with phosphorus, transferrin saturation, relative body height and vitamin C, as the major predictors.

Conclusion

In conclusion, the data presented suggest that mild to moderate vitamin and mineral deficiencies may affect physiological functions such as exercise performance, immune response, behaviour and cognitive functions. Of the single nutrients, vitamin A, vitamins of the B-complex, vitamin C and iron seem to be associated with physical working capacity and, with the addition of zinc, with immunity. In regard to cognitive function the clearest association was found with iron and some B-complex vitamins. Still, it is difficult to assess the specific effect and the relative importance of a single nutrient on the functional status of the body because in most cases the existing association was studied by univariate analysis. Since, however, some of the biochemical parameters of vitamin and mineral nutrition status are intercorrelated, the significance of correlation with the dependent variable can be reduced, or even lost in the multivariate analysis. It should also be mentioned that because of the inherently limited sensitivity of the functional tests as specific indices of micronutrient deficiency, it may prove difficult to achieve consistently reproducible effects between studies of a particular marginal deficiency even when an identical protocol is used.

Furthermore, the correlation between functional tests and micronutrient nutritional status usually increases when other micronutrients which are in inadequate supply are taken into consideration. Since, under natural conditions, inadequate nutrition often results in multiple micronutrient deficiencies, it would be more appropriate to include in the analysis all nutrients which may affect functional status of the body. In such a way we have found that about 20%–40% of the variability in physical fitness and immune function could be explained by the biochemical parameters of nutritional status. These data therefore suggest that even in the absence of precise information about the specificity and magnitude of the effects of a single micronutrient deficiency on the functional status of the body, an optimal daily micronutrient intake seems to be desirable.

References

Aebi H, Blumenthal A, Bohren-Hoerni M, Brubacher G, Frey U, Müller H-R, Ritzel G, Stransky M (eds) (1984) Zweiter Schweizerischer Ernährungsbericht. Hans Huber, Bern

Anderson R, Oosthuizen R, Maritz R, Theron A, Van Rensburg AJ (1980) The effect of increasing weekly doses of ascorbate on certain cellular and humoral immune functions in normal volunteers. Am J Clin Nutr 33:71–76

Bailey LB, Mahan CS, Dimperio D (1980) Folacin and iron status in low-income pregnant adolescents and mature women. Am J Clin Nutr 33:1997–2001

Beisel WR (1982) Single nutrients and immunity Am J Clin Nutr 35:417–468

Bidlack WR, Hamilton Smith C, Clements RA, Omaye ST (1986) Nutrition and the elderly: a scientific status summary by the Institute of Food Technologists Expert Panel on Food Safety and Nutrition. Food Technology (September), pp 81–88

Birkbeck JA (1985) Nutritional analysis of food served at McDonald's Family Restaurants, New Zealand. New Zealand Nutrition Foundation, Wellington

Bouillon RR, Auwerx JH, Lissens WD, Pelemans WK (1987) Vitamin D status in the elderly: seasonal substrate deficiency causes 1,25-dihydroxycholecalciferol deficiency. Am J Clin Nutr 45:755–763

Bowering J, Lowenberg RL, Morrison MA (1980) Nutritional status of pregnant women in East Harlem. Am J Clin Nutr 33:1987–1996

Chandra R, Newberne PM (1977) Nutrition, immunity and infection-mechanisms of interactions. Plenum Press, New York

Chandra RK, Singh G (1987) Effect of marginal and moderate malnutrition on immune function. Paper presented at the meeting on functional consequences of marginal malnutrition. WHO, Geneva

Chome J, Paul T, Pudel V (1984) Testpsychologische Untersuchung bei älteren Menschen mit Subklinischen Vitaminmangel. Ernährungs-Umschau 31:12–16

Contaldo F, Di Biase G, Lanzetta R, Lanzilli A, Simonetti MJ, Fidanza F (1987) Nutritional status of elderly people in Naples. Int J Vit Nutr Res 57:193–202

Department of Health and Social Security (1979) Nutrition and health in old age: the cross-sectional analysis of the findings of a survey made in 1972–1973 of elderly people. Reports on health and social subjects no. 16. HMSO, London

Ernährungsbericht (1984) Deutsche Gesellschaft für Ernährung. e.V., Frankfurt

ESVITAF (1986): Vitamin status in three groups of French adults. Ann Nutr Metab 30 (suppl 1)

Fidanza F, Simonetti MJ, Cucchia LM, Balucca GG, Losito G (1984) Nutritional status of the elderly. Int J Vit Nutr Res 54:75–90

Fisher S, Hendricks DG, Mahoney AW (1978) Nutrition assessment of senior Utahns by biochemical and physical measurements. Am J Clin Nutr 31:667–672

Flint DM, Wahlquist ML, Prinsley DM, Parish A, Fazio V, Peters K, Richards B (1979) The nutritional assessment of community and institutionalized elderly in Australia. Food and Nutr: Notes and Rev 36: 173–179

Goodwin JS, Goodwin JM, Garry PJ (1983) Association between nutritional status and cognitive functioning in a health elderly population JAMA 249:1917–1921

Hagman V, Bruce A, Persson L-A, Samuelsson G, Sjölin S (1986) Food habits and nutrient intake in childhood in relation to health and socio-economic conditions. A Swedish multicentre study 1980–81. VarFöche 38 (Suppl 5)

Hodges RE, Bean WB, Ohlson MA, Bleiler RF (1962) Factors affecting human antibody response. IV. Pyridoxine deficiency. Am J Clin Nutr 11:180–186

Jacob M, Hunt IF, Diriger D, Swendseid ME (1976) Biochemical assessment of the nutritional status of low-income pregnant women of Mexican descent. Am J Clin Nutr 29:650–656

Keith R, Mossholder S (1986) Ascorbic acid status of smoking and non smoking adolescent females. Int J Vit Nutr Res 56:363–366

Kerr GR, Lee ES, Lam MKM, Lorimor RJ, Randall E, Forthofer RN, Davis NA, Magnetti SM (1982) Relationships between dietary and biochemical measures of nutritional status in HANES I. Data. Am J Clin Nutr 35:214–308

McGandy RB, Russel RM, Hartz SC, Jacob RA, Tannenbaum S, Peters H, Sahyoun N, Otradovec CL (1986) Nutritional status survey of healthy non-institutionalized elderly. Energy and nutrient intakes from three-day diet records and nutrition supplements. Nutr Res 6:785–798

Mareschi JP, Cousin F, Villeon B del la, Brubacher GB (1984) Valeur calorique de l'alimentation et converture des apports nutritionnels conseillés en vitamines de l'homme adulte. Ann Nutr Metab 28:11–23
Mora JO, de Parade B, Wagner M, de Navaro L, Suescum J, Christiansen N, Herrera MG (1979) Nutritional supplementation and the outcome of pregnancy I. Birthweight. Am J Clin Nutr 32:455–462
Mc Murray DN, Loomis SA, Casazza LJ, Rey H, Miranda R (1981) Development of impaired cell mediated immunity in mild and moderate malnutrition. Am J Clin Nutr 34:68–77
Omdahl JL, Garry PJ, Humsaker LA, Hunt WC, Goodwin JJ (1982) Nutritional status in a healthy elderly population: vitamin D. Am J Clin Nutr 36:1225–1233
Pollitt E, Sacco-Pollitt C, Leibel RL, Viteri FE (1986) Iron deficiency and behavioural development in infants and preschool children. Am J Clin Nutr 43:555–565
Rush D, Stein Z, Susser M (1980) Diet in pregnancy: a randomized controlled trial of prenatal nutrition supplementation. Alan R Liss, New York
Sandstead HM (1986) A brief history of the influence of trace elements on brain function. Am J Clin Nutr 43:293–298
Schuster K, Bailey LB, Mahan CS (1981) Vitamin B6 status of low-income adolescent and adult pregnant women and the condition of their infant at birth. Am J Clin Nutr 34:1731–1735
Spurr GB (1987) Functional consequences of chronic dietary imbalance. Paper presented at the meeting on functional consequences of marginal malnutrition. WHO, Geneva
Steen B, Isaksson B, Swanborg A (1977) Intake of energy and nutrients and meal habits in 70-year-old males and females in Gothenburg, Sweden: a population study. Acta Med Scand (suppl 611), 39
Subotičanec K, Stavljenić A, Bilić-Pešić L, Gorajšćan M, Gorajšćan D, Brubacher G, Buzina R (1989). Nutritional status, grip strength and immune function in institutionalized elderly. Int J Vit Nutr Res 59:20–28
Suter PM, Russel RH (1987) Vitamin requirements of the elderly. Am J Clin Nutr 45:501–512
US Department of Agriculture (1980) Nationwide food consumption survey 1977–78. Washington DC
Van der Beek EJ (1986) Restricted vitamin intake and physical performance of military personnel. In: Predicting decrements in military performance due to inadequate nutrition. NRC. National Academy Press, Washington DC
Van der Beek EJ, van den Berg H, Schrijver J, van Dokkum W, Egger RJ (1987) Subclinical vitamin deficiency and functional performance in men. Paper presented at the meeting on functional consequences of marginal malnutrition, Geneva
van den Berg H, Schreurs WHR, Jorsten GPA (1978) Evaluation of the vitamin status in pregnancy. Int J Vit Nutr Res 48:12–21
Vir SC, Love AHG (1978) Vitamin D status of elderly at home and institutionalized in hospital. Int J Cit Nutr Res 48:123–130

Chapter 7

The Functional Significance of Micronutrient Deficiency: Vitamin C

A. Kallner

The ultimate symptom of vitamin C deficiency is scurvy, which is lethal if not adequately and promptly treated. Man is one of the few animal species which is unable to synthesize its own ascorbic acid, the antiscorbutic agent. Among mammals, only the guinea-pig, flying mammals and primates lack this ability. The metabolic deficiency is the same: lack of L-gulono-gamma-lactone oxidase, the last enzymatic step in the biosynthesis of ascorbate from glucose. Interestingly, ascorbic acid is synthesized in the liver by mammals whereas the kidney is the site of biosynthesis in other vertebrates. It is likely that species which now do not have the ability to produce ascorbate lost this ability during evolution.

Ever since its discovery in the 1920s, vitamin C has been the subject of intensive research. All aspects have been treated with varying and ever-increasing degrees of sophistication. The literature is voluminous and many good reviews have been published. One of the latest comprehensive reviews (Hornig et al. 1988) covers history, chemical and biochemical as well as nutritional aspects.

The sequence of events leading to manifest scurvy is well described in the literature. History as well as organized experiments with humans give evidence of the rate of development of the deficiency symptoms. Classical depletion studies were carried out in England during World War II (the Sheffield experiment) (Krebs 1952) and before that, Crandon et al. (1944) in Boston had studied the progress of ascorbate deficiency over 6 months. Both reports give details on the development of symptoms on diets which were designed to contain sufficient other nutrients, thus attempting to isolate the effects of ascorbate deficiency.

Briefly summarized, symptoms began to appear in some individuals after 17 weeks and all participants seemed to be affected 10 weeks later. The first signs were enlargement and keratosis of hair follicles which develop into haemorrhages with a subsequent extracellular invasion of erythrocytes. In a later stage tiny haemorrhages of the interdental papillae became evident. Other symptoms were unspecific pain and symptoms indicative of acute cardiac emergency.

Crandon et al. (1940) described the development of fatigue after about two months of depletion. This fatigue became successively more pronounced and it was also possible to document it objectively by exercise on a treadmill.

Behavioural effects of ascorbic acid deficiency have been the subject of separate studies. One of the first of these (Kinsman and Hood 1971) involved five volunteers in Iowa who were depleted for 107 days followed by a repletion phase. After as little as 1 week of depletion, one of several of the various sophisticated tests taking mental as well as physical performance into consideration indicated changed performances. It was concluded that the neurotic triad (hypochondriasis, depression and hysteria) became elevated after a few weeks of depletion, followed by personality changes and eventually reduced psychomotor performance.

One has to go back to the 1940s to find reports of widespread scurvy in society. Thus, 53 cases were reported in 1944 by McMillan and Inglis, showing very much the same signs and symptoms which were reported in the more controlled studies. Already then, this disease was characterized as a typical "bachelor's disease" the cause being ignorance of the problem, apathy or poverty.

We do see scurvy in clinical practice today also. After publishing a case report in our local medical journal (Wåger–Hörlén et al. 1986) we have been made aware of several cases annually. The reason for this apparent increase in prevalence of the disease seems to be a greater interest and awareness among physicians. There seem to have been similar experiences in other countries (Dawes and Haslock 1985).

Without exception, our cases have presented themselves with extreme fatigue, personality changes and haemorrhages in the lower limbs. Thus, the common features of these patients do not differ significantly from those observed by McMillan and Inglis in 1944. The histories of our patients disclose that it is still valid to claim that overt scurvy is a disease of the ignorant, the apathetic, and possibly of the poor and mentally disturbed.

An additional complication in these cases is multiple component deficiency which might distort the picture, and aggravate and change the symptoms. The patient, therefore, often presents himself with a complicated set of symptoms which are not necessarily recognized as deficiency symptoms. In practice patients may suffer primarily from almost anything from mental disorders to unspecific infections of the skin. Treatment frequently comprises general palliative treatment with adequate food and supervision which might lead to what is understood as spontaneous healing. As a consequence, we may not be aware of the true frequency of vitamin C deficiencies, and certainly not the frequency of the mild forms.

At this point I would like to emphasize that this experience has been gained in a western country which has a high reputation for its social welfare system. Our experience may not be applicable in other countries and other social systems.

The daily requirement of vitamin C has been the subject of lengthy discussions which are not yet concluded. As long ago as 1944 in the Sheffield study (Krebs 1953) it was shown that scurvy could be efficiently treated by giving as little as 10 mg of ascorbic acid per day. At that time, in war-time England, the authorities accepted a recommended daily intake of 30 mg to cover interindividual differences of needs. A very different approach to estimate the daily need of ascorbic acid was that introduced by Baker et al. (1969, 1971) who estimated the pool size using radioactive-labelled ascorbate and measured the daily turnover during depletion and repletion. These calculations are still used in the discussion of the recommended daily allowance of vitamin C. However, the criticism of this experimental design prompted a study based on measured and calculated kinetic parameters during steady state (Kallner et al. 1979) as concerns the vitamin C pool size. The Baker study indicated that an intake of about 45 mg would suffice to keep the estimated pool whereas the latter, kinetic study, rather pointed to an intake of about 100 mg per day considering the bioavailability and excretion rate of ascorbic acid. Recently, Olson and Hodges (1987) argued for yet other interpretations of available data.

The orthomolecular hypothesis advises an intake of 2–4 g of ascorbic acid per day (Pauling 1970). It is common knowledge that this figure was arrived at by extrapolation from the ascorbate production in rats and other mammals, based on the body weights of rats and humans. Although widely known, this hypothesis has not been accepted, probably because of lack of obvious benefits. The even higher doses advised in acute or chronic disease have failed to prove of therapeutic value in critical studies.

Nutrient density is a concept which describes the amount of micronutrient per energy unit (kcal or joule) of food intake. It now appears that this concept might bring results of orthomolecular thinking in accordance with results from direct kinetic studies in man. Thus, Newmark (1987) recently pointed out that the endogenous production of vitamin C in the rat (about 26 mg/d) and energy consumption (about 400 kcal/d) suggests an ascorbate nutrient density of 0.065 mg/kcal if there were no endogenous production. An assumed energy intake of man between 1500 and 2500 kcal would correspond to an intake of 98–163 mg ascorbate/d, compared to that of about 100 mg/d derived from the kinetic studies (Kallner et al. 1979).

Evidently, we do not yet have sufficient information to establish an optimal intake of vitamin C. Its documented non-toxicity, abundancy and claimed benefits in fighting a number of malignant and less severe conditions has led to huge daily (gram amounts) intakes by some individuals and regular intakes of amounts 10–50 times recommended doses (500–1000 mg) among many. There are no reported adverse effects of ascorbate which survive careful scrutiny but, on the other hand, there are very few, if any, well-documented cases of improved conditions besides scurvy which can be attributed to large intakes of vitamin C. A specific effect of vitamin C is improving the gastro-intestinal absorption of iron. This might very well be a presystemic effect of ascorbate and its potency in preventing sideropenic anaemia is discussed in the contribution by Leif Hallberg to this volume.

In part, the difficulty of defining an optimal intake is due to our lack of knowledge of the biochemical reactions which require ascorbic acid or are influenced by ascorbic acid. Many functions are known, and described in isolated systems, e.g., hydroxylation of proline and lysine in collagen synthesis, certain hydroxylations of steroids, biosynthesis of carnitine, which is of importance in triglyceride transportation, and oxidation of L-tyrosine in catecholamine synthesis. Ascorbic acid is also known as a general reducing agent and as a free-radical scavenger. On a functional level, vitamin C influences the absorption of iron and has been reported to be involved in the immunological and active antibacterial functions of white blood cells (Jaffé 1984). Other systemic effects are more difficult to evaluate e.g., effects on stress (Kallner 1983), serum histamine concentrations (Clementson 1980, Schorma et al. 1985) and physical performance (Gerster 1990, van der Beek 1988, Suboticanec–Buzina et al. 1984).

The systemic aspects of vitamin C and cholesterol biosynthesis and metabolism are not yet resolved.

Signs of mild deficiency include unspecific symptoms such as lassitude, fatigue, and anorexia and greater susceptibility to infection and stress. Specific symptoms occur only in advanced stages and there is an absence of easily assessable, biologically-based function tests. It seems rational and logical, therefore, to evaluate the nutritional status of vitamin C by measuring plasma or serum ascorbate. It has been shown by Kallner et al. (1979) and Hodges et al. (1969) that there exists an almost linear relationship between plasma concentration and body pools. Hodges et al. found that the first signs of scurvy appeared when the plasma concentration ranged between 1.3 and 2.4 mg/l (equal to about 7.4–13.5 μmol/l). In 1973 the National Survey of Canada defined three risk groups: high risk for adults when plasma concentration is below 10 μmol/l, moderate risk between 10 and 20 μmol/l and low risk above 20 μmol/l. If the plasma concentrations could be related to the intakes we would achieve an indication of intakes which would bring the individual into the low risk group. Such data have been accumulated by Schorah and Basu (1982). Their results are based on many studies and the variability is marked. To reach a plasma concentration of about 20 μmol/l (corresponding to the cut-off value for the low risk group) an intake of up to about 60 mg per day would be required.

At present we can, therefore, conclude that overt scurvy does not develop at intakes about 10 mg/d. Smaller intakes will cause scurvy and symptoms will appear when the serum concentration has fallen below about 10 μmol/l. Kinetic data indicate that an intake of about 100 mg/d is sufficient to reach a saturation of the ascorbate pool whereas other interpretations indicate that intakes of about 45 mg/d would be sufficient. A new approach to orthomolecular thinking would indicate intakes in the range of 100–150 mg/d. This diversity in interpretation is mirrored in the various national recommendations on daily allowances; the 40 recommendations can roughly be grouped into three levels (Table 7.1). In considering the table, one should remember that the definition of recommended allowances may differ.

Given the arguments behind the estimation of required intakes of vitamin C and the distribution of vitamin C in food-stuff, one would assume that changing the diet towards increased fibre and decreased energy intakes would increase vitamin C intake. This is because bulk food is largely derived from the plant kingdom and some sources are relatively rich in vitamin C. However, one has to bear in mind that not all vegetables are particularly rich in the vitamin, and that it is very sensitive to oxidation and thus requires careful cooking. Also, fresh vegetables and fruits are prohibitively expensive in many countries. The concept of nutrient density (Newmark 1987) might prove useful in declaring and labelling food-stuffs to guide the consumer. Nutrient density should, therefore, be considered as an alternative to the current method of nutrient concentration (amount per mass or volume).

Clearly, those who are ignorant of the necessity or, for various reasons, unable to obtain a varied, suitable diet are at risk. Traditionally, elderly and single people and young girls but also, ironically, athletes (Schorma et al. 1985) are considered risk groups. In a recently published survey of the nutritional status of the elderly in urban Boston (Jacobs et al. 1988) as many as 6% of the males and 3% of the females aged 60–98 years showed a marginal vitamin C status. Although these results may not apply to Europe or other regions, they point to

Table 7.1. RDA for vitamin C in various countries (mg ascorbic acid/day) (from Commonwealth Agricultural Bureaux (1983) 53:1–1075)

	FAO/WHO, Canada, Argentina, Finland, Hungary, United Kingdom	Scandinavia, USA, Mexico	USSR, Federal Republic of Germany, Poland, Phillipines
Infants	20	40	
Children	20	40	40
Girls	30	50	60
Youths	30	50	80
Males	30	50	75
Elderly persons	30	50	63
Pregnant women	0	+20	+90
Lactating women	0	+40	+60

an important problem. Other risk groups include those who require high intakes to reach a maximum desired pool of ascorbic acid, e.g., smokers (Kallner et al. 1981) and pregnant and lactating women. Diabetes seems to increase the metabolic turnover of ascorbate (McLennan et al. 1988) and thus predisposes to marginal deficiency.

Besides continued research on optimal vitamin C intakes, the public needs information about the importance of an adequate intake, and physicians and other health workers need education to enable them to recognize early signs of deficiency.

References

Baker EM, Hodges RE, Hood J, Sauberlich HE, March SC (1969) Metabolism of ascorbic-1-^{14}C acid in experimental human scurvy. Am J Clin Nutr 22:549–558

Baker EM, Hodges RE, Hood J, Sauberlich HE, March SC, Canham JE (1971) Metabolism of ^{14}C- and ^{3}H labelled L-ascorbic acid in human scurvy. Am J Clin Nutr 24:444–454

Clementson CAB (1980): Histamine and ascorbic acid in human blood. J Nutr 110:662–668

Crandon JM, Lund CC, Dill DB (1940) Experimental human scurvy. N Engl J Med 223:353–367

Dawes PT, Haslock I (1985) Haemartherosis due to ascorbic acid deficiency. Br J Clin Pract 39:290–293

Gerster H (1990) The role of vitamin C in athletic performance. Am J Clin Nutr. In press

Hodges RE, Baker EM, Hood J, Sauberlich HE, March SC (1969) Experimental scurvy in man. Am J Clin Nutr 22:535–548

Hornig DH, Moser U, Glatthaar BE (1988) Ascorbic acid in modern nutrition. In: Skils ME, Young VR (eds) Health and disease, 3rd edn. Lea and Febiger, Philadelphia

Jacobs RA, Otradovec CL, Russell RM et al. (1988) Vitamin C status and nutrient interactions in a healthy elderly population. Am J Clin Nutr 48:1436–1442

Jaffé GM (1984) Vitamin C. In: Machlin LJ (ed) Handbook of vitamins. Nutritional, biochemical, and clinical aspects. M Dekker, New York

Kallner A (1983) Influence of vitamin C status on the urinary excretion of catecholamines in stress. Hum Nutr Clin Nutr 37C:405–411

Kallner A, Hartmann D, Hornig D (1979) Steady state turnover and body pool of ascorbic acid in man. Am J Clin Nutr 32:530–539

Kallner A, Hartmann D, Hornig D (1981) On the requirements of ascorbic acid in man: steady state turnover and body pool in smokers. Am J Clin Nutr 34:1347–1355

Kinsman RA, Hood J (1971) Some behavioural effects of ascorbic acid deficiency. J Clin Nutr 24:455–464

Krebs HA (1953) The Sheffield experiment on the vitamin C requirement of human adults. Proc Nutr Soc 12:237–246

McLennan S, Yke DK, Fisher E, Copogreco C, Hefferman S, Ross GR, Turtle JR (1988) Deficiency of ascorbic acid in experimental diabetes. Relationship with collagen and polyol pathway abnormalities. Diabetes 37:359–361

McMillan RB, Inglis JC (1944) Scurvy: A survey of fifty-three cases. Br Med J 233–236

Nutrition Canada National Survey (1973) Information Canada, Ontario, Canada

Newmark HL (1987) Nutrient density: an important and useful tool for laboratory animal studies. Carcinogenesis 8:871–873

Olson JA, Hodges RE (1987) Recommended dietary intakes (RDI) of vitamin C in humans. Am J Clin Nutr 45:693–703

Pauling L (1970) Vitamin C and the common cold. WH Freeman, San Francisco

Salmenperä L (1984) Vitamin C during prolonged lactation: optimal in infants while marginal in some methods. Am J Clin Nutr 40:1050–1056

Schorah CJ, Basu TK (ed) (1982) Vitamin C in health and disease. Croom Helm, London

Schorma SC, Walzman M, Bannar J, Mollay A (1985) Blood ascorbic acid and histamine levels in patients with placental bleeding. Hum Nutr Clin Nutr 39C, 233–238

Suboticanec-Buzina K, Buzina R, Brubacher G, Sapunar J, Christeller S (1984) Vitamin C status and physical working capacity in adolescents. Int J Vit Nutr Res 55–60

van der Beek EJ, van Dokkum W, Schijner J et al. (1988) Thiamin, riboflavin, and vitamins B6 and C: impact of combined restricted intake on functional performance in man. Am J Clin Nutr 48:1451–1462

Wåger-Hörlén K, Kallner A, Larsson Å, Nygren AC (1986) Scurvy – a forgotten disease in modern welfare. Läkartidningen 3527–3528

Chapter 8

The Functional Significance of Marginal Micronutrient Status: Folate

K. Pietrzik

Introduction

Folate deficiency is a common problem even in industrialized countries. According to the nutrition report of the Federal Republic of Germany, folate seems to be the most critical vitamin (Deutsche Gesellschaft für Ernährung 1984, 1988) and one of the reasons for its insufficient supply might be seen in our modern lifestyles.

An adequate vitamin intake depends to a great extent on the preservation of nutrients in manufactured foods. But there are not only changes in the nutrient content of processed foods there are also changes in consumption patterns. Products that were previously seasonal may now be obtained throughout the year and new products continually appear on the market. Our food consumption changes with our changing lifestyle. These changes may be small from one year to the next, but over a longer period they become important and result in new nutritional problems.

Probability of Micronutrient Deficiency

The diet of European countries is characterized in several ways. On the one hand, we have overconsumption: on the other, we have a reduced need for energy because of decreasing physical activity and also a relatively high intake of refined foods. According to Brubacher (1978) an average individual with a daily diet of 2400 kcal (10 MJ) might eat about 90 g sugar, 30–40 g white flour, rice and so on and drink about half a litre of beer or two or three glasses of wine. Under these conditions there are only 1100–1300 kcal (4.6–5.4 MJ) left in which all necessary nutrients besides energy have to be incorporated and so it becomes difficult to provide a nutritionally adequate diet.

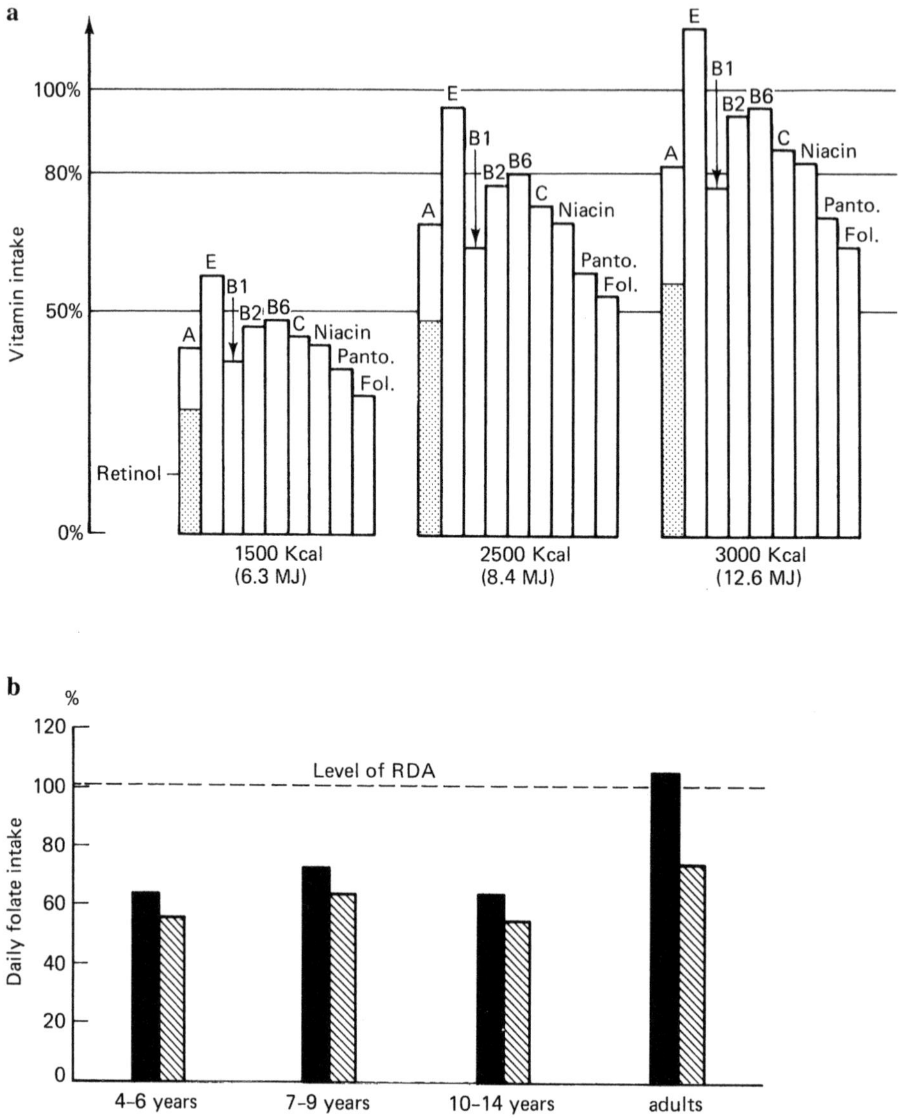

Fig. 8.1. **a** Vitamin intakes at three caloric levels based on French food habits (100% level, recommended dietary allowances). Modified from Mareschi, et al. 1984. **b** Daily folate intake as percentage of RDA (400 μg) (data calculated according to Deutsche Gesellschaft für Ernährung (1984). Filled blocks, males; hatched blocks, females.

Furthermore, many people limit their energy intake to control body weight. In this context Mareschi et al. (1984) tested diets of different energy levels (1500, 2500, 3000 kcal). The vitamin contents of the diets were calculated from food consumption tables taking into account losses resulting from usual cooking methods.

The results show that there is a high risk that the supply of some vitamins, especially folic acid, may be inadequate (Fig. 8.1a). This low folate supply with lower calorie intake is not surprising as the folate content of a so-called ideal diet (containing vegetable, fruits etc.) will just reach the RDA, but only in male adults (calculated on German data according to the Deutsche Gesellschaft für Ernährung (1985). In females, the threshold of 80% of the RDA is just reached. In the case of children the supply situation is worse since their folate requirement is the same as for adults but their energy needs are less (Fig. 8.1b). From these data there is clear evidence that a low-energy diet cannot guarantee an adequate folate intake. The results of epidemiological surveys made in other European countries are in agreement with the results mentioned above. The threshold of 80% of the recommended allowances in general is not reached with an energy intake of 2500 kcal (6 MJ). With an intake of 1500 kcal (3.5 MJ) most of the vitamins do not (or just) reach 50% of the recommendation and folate is only about 30% of the RDA. People with an energy intake on this level – or even less – over a long period are recommended to take folate supplements or vitaminized food in order to fulfill the daily requirements.

Another reason for an inadequate vitamin supply results from processing and home cooking as vitamins are not very stable. The most sensitive vitamins are ascorbic acid and folate. Their maximal losses can be up to 100% depending on the exposure to the different agents. Cooking for two minutes, for example, destroys 80% of the folate in vegetables. With longer cooking and warm holding of meals there is complete destruction of folate. Thermal processing, exposure to light and oxygen are the most harmful agents for folate (Fig. 8.2). Although food has been processed since ancient times, modern preservation and processing techniques such as canning and freezing are of relatively recent origin. Their advantages and disadvantages may be summarized as follows. Modern food processing techniques are necessary to provide year-round delivery of wholesome, nutritious and attractive food. On the other hand, nutritional losses take place – especially of water soluble vitamins, which are damaged by various processing and preservation techniques both industrially and at home. Determinations of some of the vitamins, especially folate, are so inaccurate that comparisons between different investigations are questionable. To calculate the contribution of folate from food it is essential to know the content of the various

	pH7	<pH7	>pH7	oxygen	light	temperature	maximum losses %
vitamin A	●	↓	●	↓	↓	↓	40
vitamin B1	↓	●	↓	↓	●	↓	80
vitamin B2	●	●	↓	●	↓	↓	75
vitamin B6	●	●	●	●	↓	↓	40
vitamin B 12	●	●	●	↓	↓	●	10
vitamin C	↓	●	↓	↓	↓	↓	100
vitamin D	●		↓	↓	↓	↓	40
vitamin E	●	●	●	↓	↓	↓	55
vitamin K	●	↓	↓	●	↓	●	5
biotin	●	●	●	●	●	↓	60
folate	↓	↓	●	↓	↓	↓	100
pantothenic acid	●	↓	↓	●	●	↓	50

Fig. 8.2. Potentially deleterious agents for vitamins and maximum losses. A *circle* indicates that the vitamin is stable to the agent and an *arrow* that it is not.

forms found in the food. Folate occurs in food as a mixture of pteroylpolyglutamic and pteroylmonoglutamic compounds, and 80% of the absorption is from pteroylmonoglutamate; polyglutamic compounds are less well absorbed. When we do not know the proportion of the different compounds there is no reliable basis for the exact calculation of absorption.

Sequence of Events in Folate Deficiency

The pioneering studies of Victor Herbert in the 1960s revealed the sequence of events in the development of symptoms of folate deficiency (Herbert 1962, 1964, 1965). With virtually folate-free diets (<5 μg/day), blood folate concentrations fall within two weeks followed after about 17 weeks by the development of hypersegmentation of the polymorphonuclear granulocytes (Fig. 8.3).

Soon after this symptom FIGLU excretion is increased, followed by diminished erythrocyte folate concentration (16–18 weeks). Folate concentrations in red blood cells reflect body stores and when these are depleted, haematopoesis follows, characterized by megaloblastic bone marrow and followed by megaloblastic anaemia.

Megaloblastosis is very seldom found in industrialized countries. Only chronic alcoholics have a high prevalence of severe folate deficiency and nearly 50% of them have megaloblastic anaemia. This high prevalence is not only due to folate deficiency in the diet but also to interactions between alcohol and folate which reduce absorption (Hillman and Steinberg 1982, Steinberg et al. 1981).

The increased requirements during pregnancy explain why up to 5% of pregnant women develop megaloblastic bone marrow characterized by proerythrocytes and megaloblasts. A smaller number develop megaloblastic anaemia characterized by a diminished red cell content in the peripheral blood, and occasionally these megaloblasts can be detected in the circulation, whereas normally they are found only in the bone marrow (Jaud 1979, Shonjania 1984).

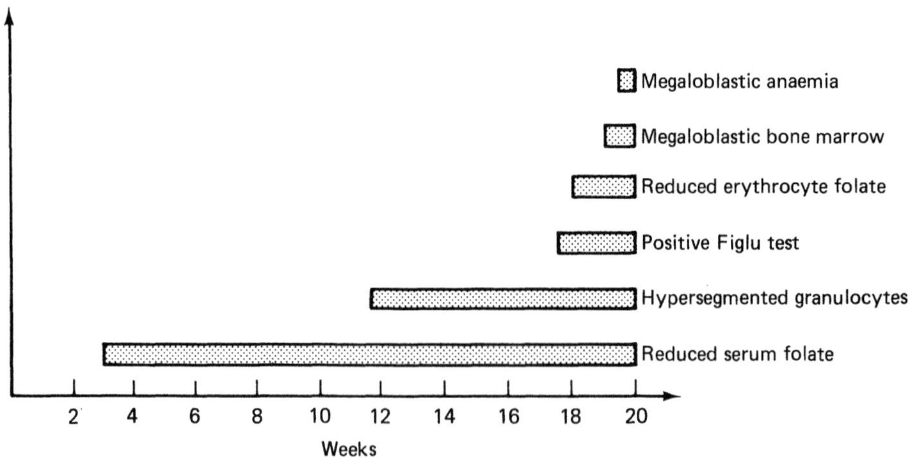

Fig. 8.3. Sequence of events in the development of folate deficiency symptoms (Herbert 1962).

The functional consequences of megaloblastic anaemia are on aerobic capacity, workload and disease resistance. These are very well documented in the literature and need not be discussed here.

Haematological Findings in Marginal Folate Deficiency

Basing on Herbert's findings on hypersegmentation as an early sign of folate deficiency our group attempted to correlate serum and erythrocyte folate levels with alterations in polymorphonuclear granulocytes. It was hoped that careful observation of peripheral blood smears for hypersegmentation would allow the early detection of folate deficiency long before the development of the more severe later changes, and it was found that groups with relatively high mean values of serum folate levels (>20 ng/ml, >10 ng/ml and >4.5 ng/ml) do not differ significantly in the segmentation of the neutrophilic granulocytes. The mean lobe average of these well nourished (folate) groups was 3.3 (Fig 8.4).

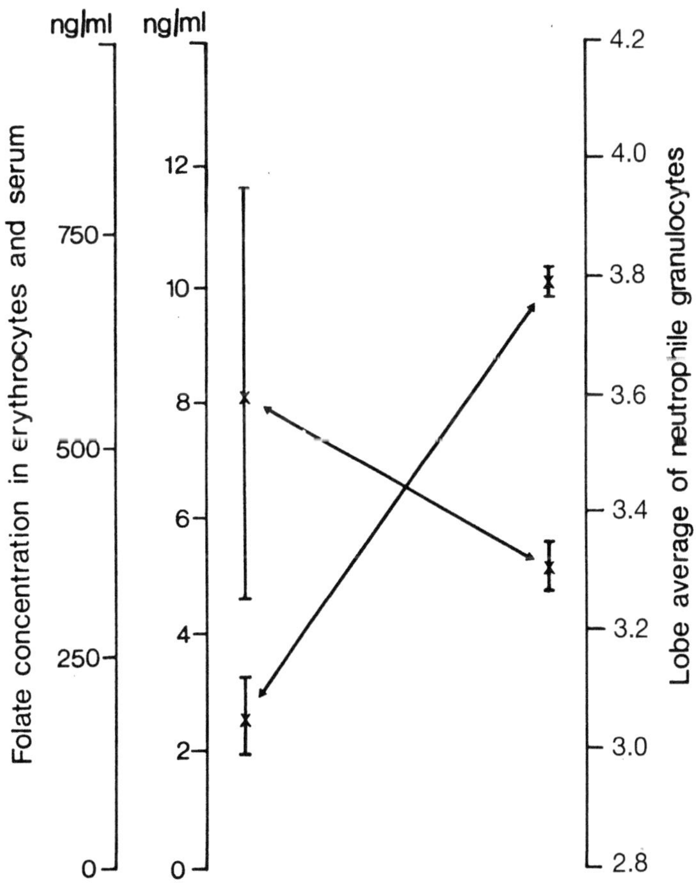

Fig. 8.4. Folate tissue concentration in relation to lobe average of polymorphonuclear granyulocytes.

a

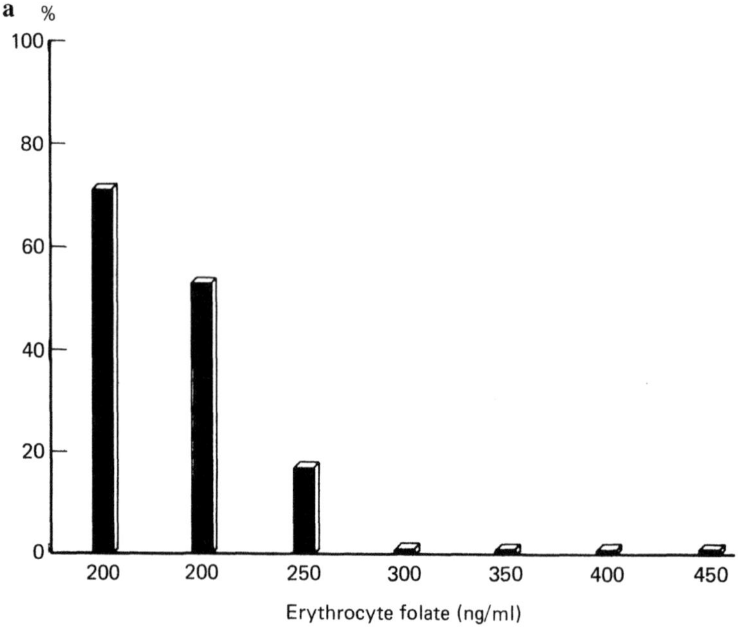

b

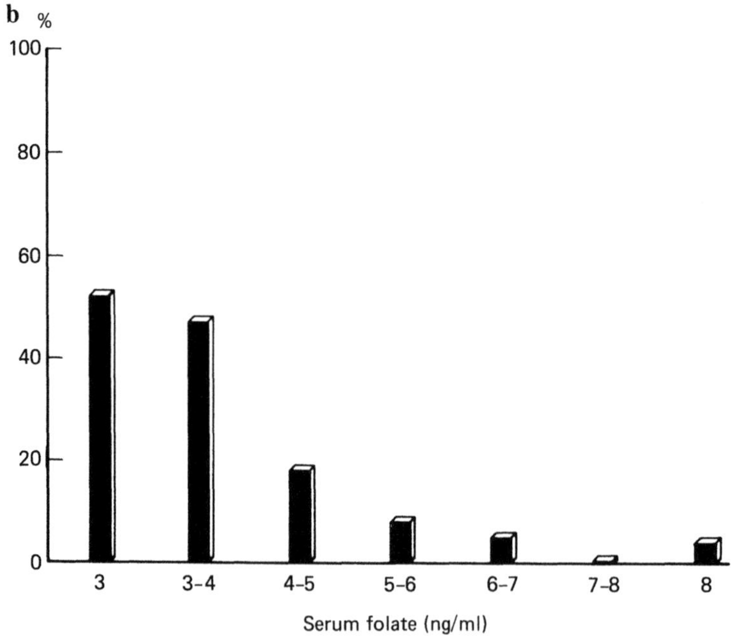

Fig. 8.5a,b. Frequency of hypersegmentation (lobe average >3.6) of polymorphonuclear granulocytes in relation to, **a**, erythrocytic and, **b**, serum folate levels.

On the other hand, serum levels <4.5 ng/ml correlate with an increasing lobe average, indicative of an inadequate vitamin intake. Serum levels <3.5 ng/ml differ highly significantly in hypersegmentation with a lobe average increasing up to >3.6.

According to these findings a serum folate level <4.5 ng/ml can be regarded as an early sign in the development of folate deficiency and serum levels lower than 3.5 ng/ml reflect a more severe situation (Pietrzik et al. 1978a, Hages et al. 1987). In order to verify these criteria for the evaluation of the nutritional status of folate, we additionally analysed the erythrocyte folate concentrations in different groups of volunteers.

With decreasing erythrocyte folate levels there is a close correlation with an increasing lobe average. Erythrocyte folate concentrations of <250 ng/ml correspond highly significant with hypersegmentation of neutrophilic granulocytes exceeding 3.6 lobes in the average. We therefore suggest that the red cell folate levels of <250 ng/ml as well as hypersegmentation (>3.6) are indicative of a chronically insufficient folate supply with a high risk of developing megaloblastosis (Pietrzik et al. 1978b).

These correlations are shown in Fig. 8.5, which indicates the frequency of hypersegmentation (>3.6 segments) in relationship to serum resp. erythrocyte folate concentrations. With high folate levels e.g. 20, 10 or 5 ng/ml in serum virtually no neutrophilic granulocytes are found with a lobe average exceeding 3.2. With decreasing vitamin levels the number of hypersegmented granulocytes increases and there is a highly significant increase if serum levels are 3.5 ng/ml or less. The same relationship exists between the erythrocyte folate levels and lobe average of neutrophils. Decreasing erythrocyte folate levels, <250 ng/ml, are closely correlated with an increase in hypersegmentation of polymorphonuclear granulocytes (lobe average >3.6) (Pietrzik 1986).

These morphological changes correlate highly significantly with folate, as vitamin B12 deficiency can be excluded in our studies.

Biological Importance and Frequency of Marginal Folate Status

To clarify the biological importance of these cell changes we looked for the effect of a folate-loading test on the maturation of neutrophils and the formation of the nucleus (Fig. 8.6). In this study all the volunteers took 1000 μg folate daily for two weeks. All volunteers with normal or high serum levels and low lobe average showed no influence on cell maturation or on the formation (lobe average) of the nucleus (data not documented in Fig. 8.6). However in cases of folate deficiency, indicated by hypersegmentation (and additionally by low serum and erythrocytic folate levels), the vitamin supplement influenced cell formation, as the lobe average tended towards normal during this relatively short test-period (Pietrzik 1986, Hages et al. 1987). Based on these biological indices nearly 15% of various population groups showed folate deficiency in a relatively advanced stage (Table 8.1). In young people (18–24 years), for example, we found 28% of the females and 19% of the males folate deficiency accompanied by hypersegmentation. We examined 420 industrial workers (20–50 years) and

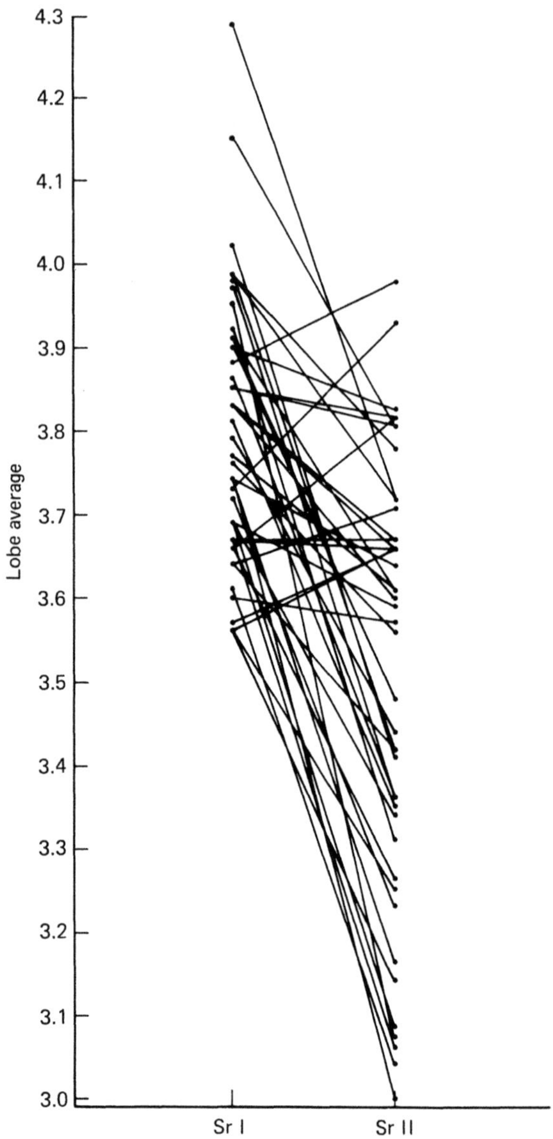

Fig. 8.6. Alteration of lobe average in hypersegmentation before (SrI) and after (SrII) an oral folate load of 1mg daily for two weeks. The figure shows that the lobe average tends towards normal.

found endemic folic deficiency in them as well as in the elderly. Younger people in their twenties – especially women – seem to be much more affected than the other adults tested.

Our recent studies in children demonstrate the importance of folate for cell maturation and development, as in the puberty period folate requirement is increased. So it was not surprising to find that children in the age group of 11–15 years showed a significantly higher prevalence of folate deficiency, 29.1%,

Table 8.1. Folate status in different groups of people in the Federal Republic of Germany (from Deutsche Gesellschaft für Ernährung 1988). The figures are for the percent of the population listed with low folate status

		Folate status %	
	Age (years)	Males	Females
Adults	18–24	19.0	28.0
	20–50	12.0	17.0
	65–90	15.5	13.5
Children	1–5	10.0	
	6–12	13.7	
	13–16	29.1	

compared with 10% and 13.7% in the younger children (Tab. 8.1) (Hages et al. 1986).

Functional Significance of Marginal Folate Deficiency

Basing on Herbert's findings we examined the functional consequences of hypersegmentation. The biochemical reason for an increasing lobe average is folate deficiency as formylfolate is involved in purine synthesis (Fig. 8.7).

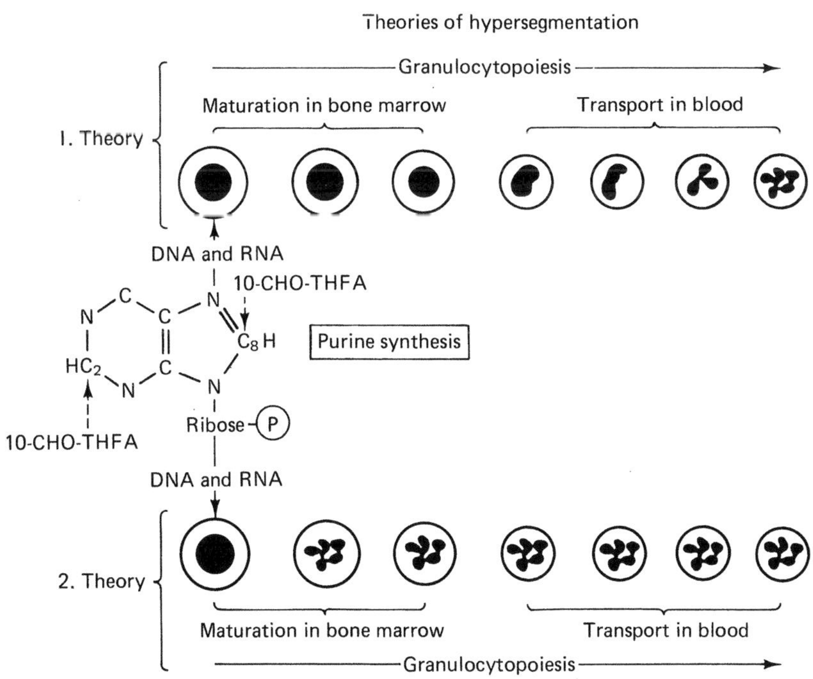

Fig. 8.7. Theories of hypersegmentation.

The C-atoms in positions 2 and 8 of the purine bases originate from folate metabolism and these purines and pyrimidines are essential for DNA and RNA formation. In cases of severe folate deficiency cell proliferation is affected and a diminished number of red and white cells are found in the circulation, combined with megaloblastic findings as described above. When folate deficiency is mild, only the nucleus of the granulocytes is affected although details are not known. According to one theory, segmentation is due to ageing of the normal granulocytes. There is an insufficient renewal of granulocytes and the body tries to maintain the number of granulocytes in the circulation; during this period the formation of the nucleus becomes more and more hypersegmented (Lindenbaum and Nath 1980). An alternative theory is that there is an initial DNA abnormality in the bone marrow, because of folate deficiency (Fig. 8.6) (Boll 1976, Nath and Lindenbaum 1979).

There is little information of the functional significance of marginal folate status. Despite several investigations of nutritional status and neutrophilic function there is no clear evidence for the role of folate. For example, in one study (Youinou et al. 1987) 92 people suffering from protein-calorie malnutrition were examined for neutrophilic function and its relation to folate status. Serum-folate levels were above 3 ng/ml in 38 patients and 3 ng or less in 54 patients. Significant differences were found between those two groups of patients with regard to phagocytosis (Youinou et al. 1982). However these studies were carried out with people suffering from protein energy malnutrition, which itself influences the immunological status in general and phagocytosis in particular.

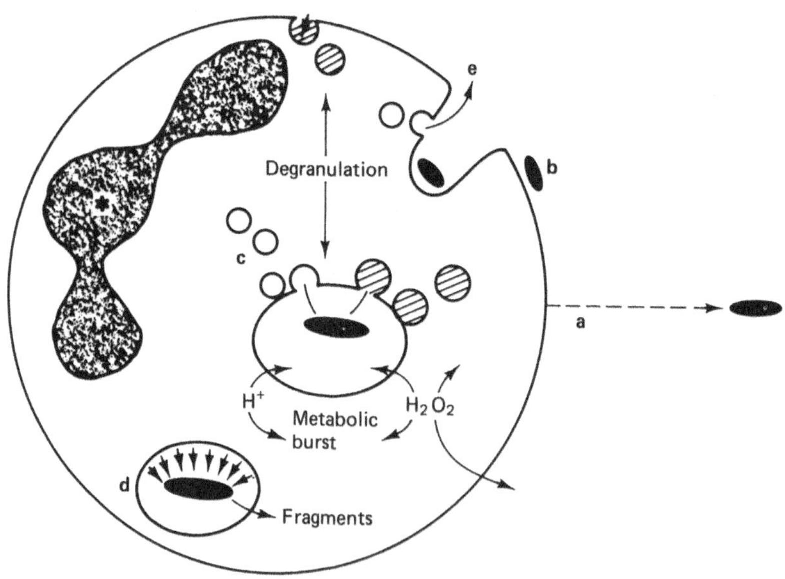

Fig. 8.8. Aspects of the neutrophil response concerning chemotaxis **a**, phagocytosis **b**, intracellular killing **c**, digestion **d**, extracellular release **e**. Asterisk marks cell nucleus.

The possible link between folate status and polymorphonuclear function remains to be elucidated. In this context we have started a study examining phagocytic activity of granulocytes in folate-deficient volunteers, keeping all other nutrients at an optimal status. In one approach we count, under the microscope, the number of particles ingested by neutrophils and correlate them with folate status and lobe average (Fig. 8.8). Another method deals with the chemotactic activity of neutrophils, using Boyden chambers; finally, we look for disorders of peroxidation in the granulocytes by measuring oxygen and hydrogen peroxide production, needed to destroy the ingested bacteria.

Though our first findings support the results of Youinou we are not yet able to give a definite answer concerning the functional consequences of hypersegmentation and marginal folate status.

References

Boll I (1976) Granulozytopoese – Morphologie, Physiologie, Kinetik und Funktion. In: Begemann H (ed) Handbuch der inneren Medizin, vol II, part 3: Leucocytäres und reticolocytäres System. Springer-Verlag, Berlin Heidelberg New York

Brubacher GB (1978) Importance of meat products for the supply of micronutrients in human nutrition. In: Downey WK (ed) Food quality and nutrition research priorities for thermal processing. Applied Science Publishers, London, pp 462–467

Deutsche Gesellschaft für Ernährung e.V. (DGE) Ernährungsbericht (1984) Hrsg. DGE im Auftrag des Bundesministers für Jugend, Familie und Gesundheit und des Bundesministers für Ernährung, Landwirtschaft und Forsten

Deutsche Gesellschaft für Ernährung (DGE) (1985) Empfehlungen zur Nährstoffzufuhr. Umschau Verlag, Frankfurt

Deutsche Gesellschaft für Ernährung (DGE) Ernährungsbericht (1988) Hrsg. DGE im Auftrag des Bundesministers für Jugend, Familie und Gesundheit und des Bundesministers für Ernährung, Landwirtschaft und Forsten

Hages M, Mirgel C, Pietrzik K (1987) Folsäure: ein kritisches Vitamin. Eine Übersicht zum aktuellen Stand der Folatforschung. Vit Min Spur 2:155–169

Hages M, Pietrzik K, Rotthauwe HW, Weber HP, von Schnakenburg K (1986) Zur Folatversorgungssituation bei Kindern Sozialpäd. Praxis und Klinik 8:23–29

Herbert V (1962) Minimal daily adult folate requirement. Arch Intern Med 110:649–652

Herbert V (1964) Studies of folate deficiency in man. Proc R Soc Med 57:377–384

Herbert V (1965) Drugs effective in megaloblastic anaemias. Vitamin B12 and folic acid. In: Goodman LS, Gilman A (eds) The pharmacological basis of therapeutics, 3rd ed. Macmillan, New York, pp 1410–1442

Hillmann RS, Steinberg SE (1982) The effects of alcohol on folate metabolism. Ann Rev Med 33:345–354

Jaud W (1979) Folsäure und Eisenmangel in der Schwangerschaft. Frauenheilk 39:317–323

Lindenbaum J, Nath B (1980) Megaloblastic anaemia and hypersegmentation. Br J Haematol 44:511–513

Mareschi JP, Cousin F, Villeon B de la, Brubacher GB (1984) Valeur calorique de l'alimentation et couverture des apports nutritionnels conseillés en vitamines de l'homme adulte. Ann Nutr Metab 28:11–23

Nath B, Lindenbaum J (1979) Persistence of neutrophil hypersegmentation during granulopoiesis. Ann Intern Med 90:757–760

Pietrzik K (1986) Folsäure-Mangel. Fachgespräch am 5 Juli 1986 in Rottach-Egern. W. Zuckschwerdt Verlag, Munich

Pietrzik K, Urban G, Hötzel D (1978a) Biochemische und haematologische Maßstäbe zur Beurteilung des Folatstatus beim Menschen. I. Beziehung zwischen Serumfolat und neutrophilen Granulozyten. Intern Z Vit Nutr Res 48:391–396

Pietrzik K, Urban G, Hötzel D (1978b) Biochemische und haematologische Maßstäbe zur Beurteilung des Folatstatus beim Menschen. II. Vergleichende Messung von Folat in Serum und Erythrozyten. Intern Z Vit Nutr Res 48:397–401

Shonjania AM (1984) Folic acid and Vitamin B12 deficiency in pregnancy and in neonatal period. Clin Perinatol 11:433–459

Steinberg SE, Campbell CL, Hillmann RS (1981) The effect of alcohol on hepatic secretion of methylfolate (CH3h4PGA) into bile. Biochem Pharmacol 30: 96–98

Youinou PY, Garre MA, Menez JF, Boles JM, Morin JF, Pennec Y, Miossec PJ, Morin PP, Menn GL (1982) Folic acid deficiency and neutrophil dysfunction. Am J Med 73:652–657

Chapter 9

Vitamin A Deficiencies and Latent Deficiencies

O. Amédée-Manesme, M. S. Mourey and C. Carlier

Vitamin A is an essential nutrient for man and animals. The long-term lack of an adequate supply of vitamin A or its precursors leads to the appearance of disorders of increasing gravity. Ultimately, the subject becomes blind and then dies. Vitamin A deficiency is a major problem in public health. It is serious because it is the primary cause of blindness throughout the world; it is serious because partial or total blindness is a burden not only for the individual but also for the community; it is serious because it affects children first and foremost and is one factor increasing infant morbidity and mortality.

Metabolism and Function of Vitamin A

The term vitamin A is used for the ß-ionone derivatives which possess a structure or a biological activity comparable to that of the parent molecule, trans-retinol (or retinol). The most accurate analytical methods make use of the physico-chemical properties of the retinoids: absorption in the ultraviolet (UV) at 325 nm, fluorimetry, high performance liquid chromatography in combination with UV absorption of fluorescence measurements. The term provitamin A is used for all of the carotenoids possessing a biological activity comparable to that of vitamin A.

Vitamin A is present in foodstuffs in two forms: retinyl esters and provitamin A. The retinyl esters are found only in foodstuffs of animal origin such as liver, milk, butter, cheese, eggs and fish. The provitamins in the form of carotenoids are found in the plant kingdom. Certain plants and fruits are particularly rich in ß-carotene. Among the latter are leafy green vegetables (spinach, green cabbage, cassava leaves, amaranth), yellow-orange vegetables (carrots, squash), certain tubers such as yellow maize and yams, and yellow-orange fruits (apricots, papayas, mangoes). The concentration of preformed vitamin A in a foodstuff or meal is usually expressed in micrograms of retinol. The amount of provitamin A (carotenoids) is also expressed in micrograms. Since provitamin A is biologically such less active than retinol, and to simplify the calculation of the overall vitamin activity of a meal or a diet, it has been agreed to use the concept of retinol equivalent (RE):

1 μg of retinol = 1 μg of retinol equivalent (RE)

1 μg of ß-carotene originating in food = 0.167 μg of retinol equivalent

1 μg of other carotenoids with vitamin A activity and originating in foods = 0.084 μg of retinol equivalent (RE)

The international unit (IU) of vitamin A is equal to 0.3 μg of retinol.

The average needs of a population, taking into account the different classes according to age and sex, are about 400–500 μg RE, and are probably slightly higher in developing countries than in the industrialized countries on account of the younger demographic structure. At the global level, about 220 μg of preformed vitamin A (retinol) and 3000 μg of carotenoids are available per person per day.

Derivatives of vitamin A (mainly in the form of retinyl esters) and provitamins A (mainly in the form of ß-carotene) are supplied in the diet. The compounds of the vitamin A group are subjected to the sequential action of gastric and/or pancreatic and intestinal secretions. The retinyl esters are hydrolysed by the pancreatic retinol ester hydrolase and by a hydrolase of the brush border of the enterocyte. After this hydrolysis, the retinol is incorporated into micelles formed as a result of the action of biliary secretions and is absorbed by a probably active mechanism. In the interior of the intestinal cell, the retinol is re-esterified, mainly as the palmitate, incorporated into chylomicrons, and excreted into the lymph. It rejoins the general circulation via the thoracic canal. The liver contains 90% of the vitamin A of the organism.

In the liver, retinol is released from retinyl palmitate by the action of a retinyl ester hyrolase. It rejoins the pool of retinol molecules. It can then bind to the plasma retinol-binding protein (pRBP) in which form it passes into the plasma (the complex thus formed is called holoRBP) or can be taken up by the cytoplasmic RBP and transferred to storage sites which consist of the lipid globules of Ito cells and hepatocytes. Retinol is stored in the form of retinyl palmitate as a consequence of the action of retinol acyl transferase. In response to a stimulation of still unknown origin, the retinyl ester hydrolase releases retinol from the lipid globules. It then rejoins the pool of retinol molecules. In the liver, retinol can be oxidized to retinal, then to retinoic acid, phosphorylated to retinyl phosphate, converted to a glucuronic acid conjugate and excreted in the bile. Retinol is released into the plasma bound to plasma RBP and prealbumin. After binding to membrane receptors, retinol enters the target cell and plasma RBP is returned to the circulation, then degraded or recycled.

The functions of vitamin A are numerous.

Vitamin A plays a role in dark adaptation which is associated with the presence of rhodopsin in the rods of the retina of the eye, a photosensitive pigment synthesized from available 11-cis-retinal.

Retinoic acid possesses two specific nuclear receptors. Through this system, retinoic acid can act directly on the expression of the genome.

The results of epidemiological surveys and of experimental studies in animals and on human populations suggest that certain nutritional factors can influence the incidence of various forms of cancer. They may act by promoting carcinogenesis or by inhibiting it. Retinol and ß carotene are among the nutrients that are capable of exerting a protective effect.

Diagnosis of Vitamin A Deficiency

The evaluation of vitamin A reserves in man is a difficult problem: since 90% of the vitamin A is stored in the liver, liver biopsy is the standard method which enables the existence or absence of a vitamin A deficiency to be established. Recently, it has been demonstrated that the evaluation of the amount in the liver on the basis of a liver biopsy at least 1cm long was representative of the mean content of the liver in spite of the heterogeneity of hepatic distribution. The best method of liver analysis makes use of a high performance liquid chromatography system coupled to a spectrophotometer adapted to absorption measurements at 325 nm. According to the International Vitamin Consultative Group, a deficit is defined as less than 20 μg of vitamin A/g of liver. In practice, a classification into three groups has been suggested: deficient if less than 10 μg, marginal between 10 and 20 μg, normal if higher than 20 μg/g of liver. This normal or acceptable state in respect of vitamin A is defined in relation to the duration of protection. This is the period during which a subject can continue to function normally in spite of an insufficient intake of vitamin A. This period amounts to 100 days for an initially well-fed child weighing 15 kg.

The other methods of evaluating the vitamin A status – clinical, biological and histological – are indirect.

Clinical

Vitamin A deficiency affects the epithelial structures of most organs, the eye being the most obvious example, but injury to the epithelia of the respiratory and digestive tracts is likely to cause pulmonary and digestive disturbances.

Damage to the eye can be external involving injury of the cornea and conjunctive epithelia (xerophthalmia) or internal involving reduction of the sensitivity of the retina to light (night-blindness). A precise classification of the types of ocular damage has been suggested by WHO.

Conjunctival xerosis is due to a transformation of the epithelium of the conjunctiva from the normal cylindrical type into an epithelium of the stratified squamous type with the disappearance of mucoid cells, formation of a granular cell layer and keratinization of the surface.

Conjunctival xerosis combines several of the following characteristics often seen better in oblique lighting:

The normal brighteners of the bulbar conjunctiva give way to plaques "like sand banks as the tide goes out".

A loss of transparency of the conjunctiva in which the vascular system is masked and only the large arterioles remain visible.

A tendency to thickening and hardening of the conjunctiva: the appearance of more or less vertical small folds, which are easily demonstrated by folding the temporal border of the conjunctiva against the external canthus.

A fine, diffuse and hazy pigmentation in dark-skinned subjects.

The incidence of most of these symptoms is strongly suggestive of the diagnosis of stage X1A. This stage is difficult to diagnose and cannot be recommended as a screening criterion in a community. Bitot's spot or X1B is a superficial silvery-grey patch, often with a foamy surface, on top of the surface of the bulbar conjunctiva. This patch appears in the first place in the temporal quarter and is very often bilateral.

Corneal xerosis or X2 follows conjunctival xerosis. The first stage of the attack is punctate keratitis which starts in the lower nasal quadrant. This stage can be detected by examination by means of a slit lamp after staining with fluorescein. The X2 stage is specific when the corneal surface appears wrinkled, "gritty" and lacking sparkle on clinical examination. This stage responds to treatment within 2 to 5 days and is cured within 2 weeks.

Corneal ulceration with xerosis or X3A is the first irreversible stage in the destruction of the eye. It is characterized by a loss of substance from a part or all of the thickness of the cornea often situated in the lower half of the cornea.

Keratomalacia or stage X3B consists of a softening of the entire thickness of a part (at least a third) or more often of the entire cornea, leading invariably to deformation or destruction of the eye-ball. The condition progresses rapidly. The corneal structure disappears and is replaced by a gelatinous mass. The extrusion of the lens and the loss of vitreous material can occur. The stages X3A and B are decisive for vision since they correspond to a permanent destruction of a part or all of the corneal stroma and thus to a permanent impairment of the structure of the cornea.

Vitamin A deficiency can be accompanied by secondary symptoms:

Night-blindness or stage XH is due to a diminution in the synthesis of the rhodopsin of the rods. It can be detected by questioning the family.

The XF stage is a special feature of the fundus oculi consisting of multiple patches disseminated along the vessels. These patches in combination with night-blindness are very suggestive of xerophthalmia.

The XS stage is a consequence of the cicatrization of the irreversible corneal lesions mentioned above. It is referred to as leukoma.

Biochemical

Plasma Retinol

The plasma concentration of vitamin A only reflects the reserves of the organism when most of the hepatic reserves have been exhausted and the plasma concentration is lower than 10 μg/dl or 0.35 μmol/l. This concentration is evidence of hepatic deficiency of vitamin A. The analysis of plasma retinol is useful for the evaluation of the vitamin A status of a population. In France, the mean plasma concentration of boys (0–14 years old) is 43 μg/100 ml (n=207) and 41.8 μg/100 ml (n=185) for girls (0–14 years old). In this normal population, the distribution curve for plasma retinol ranges between 20 and 80 μg/dl. In Senegal, for example, the mean plasma concentration of children under 6 years of age is less than 20 μg/100 ml. The distribution curve for plasma retinol is, of course, markedly lower than in France with plasma concentrations varying from 1 to 60 μd/dl. In this sense the estimation of the mean plasma concentration of retinol may be useful.

Plasma Retinol Binding Protein

The concentration of plasma retinol binding protein (pRBP) does not seem to be a good criterion for assessing vitamin A deficiency since it is relatively insensitive and subject to variations from other causes such as malnutrition and inflammation. The ratio retinol/RBP, expressed in moles, may be an indication of vitamin A deficiency if it is lower than 0.6.

The Relative Dose Response Test

The relative dose response test (RDR test) is a dynamic blood test. When retinyl palmitate reaches the liver, it is hydrolysed to palmitic acid and retinol. The latter is either stored or bound by the pRBP and released into the general circulation. In the case of vitamin A deficiency, the plasma level of retinol lies below the homeostatic threshold of the subject. This leads to retinol being released so that the plasma concentration again approaches the homeostatic level. If vitamin A is not deficient, the retinol is stored. If the concentrations of plasma retinol are compared at times 0 (before retinyl palmitate is supplied) and 5 h afterwards, an increase in plasma retinol is observed which is greater the lower the level of vitamin A in the liver. The RDR test can be expressed by the following formula:

$$\frac{\textit{plasma retinol (T 5 h)} - \textit{plasma retinol (T 0)}}{\text{plasma retinol (T 5 h)}} \times 100 = \mathrm{X}\%$$

The retinyl palmitate may be supplied orally or intravenously in cases of malabsorption. In the normal adult, a result higher than 15% corresponds to a vitamin A deficiency (hepatic levels less than 20 μg/g liver); in the healthy child or one suffering from liver disease, a RDR test performed intravenously and giving a result higher than 20% provides evidence of vitamin A deficiency. If the result of the RDR test is lower than 10%, vitamin A status is normal. The RDR test is a sufficiently accurate criterion for assessing reserves of vitamin A.

Histological or Cytological Impression

The ocular impression test is based on the cellular differentiation induced by vitamin A. The transfer technique involved in the ocular impression test makes use of cellulose acetate filter paper for taking samples of cells from the conjunctiva of each eye.

Millipore filter paper (HAWP) is cut into strips of 5 × 25 mm; the entire width of about 5 mm of one end of a strip is placed on the lower temporal quadrant of the bulbar conjunctiva for 3 to 5 seconds. The paper is withdrawn without it touching the eye. It is placed immediately on a glass slide; the cells thus collected are transferred by simply pressing the paper to the slide with the fingers. The cells are fixed for 15 min in 95% ethanol, the slide is dried in air then immersed in the stain for 15 – 20 min; after being washed with tap water, the sample is examined under the microscope. The stain is composed of one volume of carbolfuchsin (Ziehl–Nielsen solution) and two volumes of 0.2% Alcian blue in 5% acetic acid; this stain is stable at ambient temperature. It is possible to use a combined fixation and staining bath by preparing a mixture of one volume of ethanol and 8 volumes of stain. When the conjunctival epithelium is normal (hepatic vitamin A greater than 20 μg/g liver), the cells in the sample are of two types:

A mass of very many pink epithelial cells, with dark pink nuclei; the boundaries of the membranes are invisible or scarcely visible and the cytoplasm/nucleus ratio is about 1.5.

Fewer cells stained blue with red nuclei: these are mucoid cells or mucocytes or "goblet cells", their density is variable and they are distributed in an irregular manner throughout the conjunctiva, hence the need to take the sample always from the same area of the eye-ball. Traces of mucus stained blue are also observed outside the cells, reflecting either free mucus on the conjunctiva or mucus derived from mucocyctes which burst when the sample was taken. The subjects are considered to be normal or N.

In the case of vitamin A deficiency, the imprint shows that the epithelial cells are far fewer, isolated and clustered in small groups; the cytoplasms are large (× 3 or 4) and the nuclei are small, even pyknotic; there are no mucocytes. The subjects are considered to be deficient or D.

All of the intermediate stages between these two extremes, normal and deficient, can be observed, epithelial cells becoming larger and more isolated as the size of the nucleus diminishes and mucocytes become rarer. It is not known how to interpret the significance of these forms. They may be classified as forms showing a risk of deficiency. The subjects in these catagories are considered to be marginal or M. Subgroups have been proposed in this category.

An excellent correlation exists between the appearance of deficiency given by the ocular impression test and the amount of hepatic vitamin A below 20 μg/g: this test provides an estimate of a preclinical deficiency.

This test has the advantage of being non-invasive, reliable and inexpensive; it can be performed anywhere with the minimum of equipment and the result can be read off very easily. It seems to be quite well adapted to mass screening, particularly in developing countries. It is important to emphasize the limitations of the test:

The difficulty of taking a sample from very young children; this test is very difficult to perform on children under 2 years of age (10–15% failure rate); it is impossible to obtain a useful sample if the child cries.

It is necessary to take the sample from the inferior temporal area. The cells of the palpebral fissure may have an intermediate or deficient appearance (keratinization).

There are technical problems associated with its performance in the field:

The impact of ophthalmological diseases. The ocular impressions test is falsified by trachoma, the conjunctivitises, tropical lmbo-conjunctivitis (study conducted on 1500 subjects). The test cannot be used when the subject is suffering from an ophthalmological disease.

The specificity and sensitivity of this test must be defined in relation to the exact clinical state of the patients.

Recent studies conducted at Djibouti, in Senegal, the Congo, Cameroon and France have made it possible to alleviate most of these problems:

The quality of the sample and its localization can be checked by a test using Rose Bengal dye but this test is painful.

When it is very hot, the test may be performed in the morning or in the evening.

Of all the techniques that can be used at present for the evaluation of vitamin A status, the transferred ocular impression seems to us to be a method which can be used in all countries, and can thus make it possible to fight effectively against the ravages of hypovitaminosis A. It deserves further study in order to improve the definition of the intermediate forms, the influence of malnutrition and the impact of ophthalmological diseases.

Vitamin A Deficiency

Vitamin A deficiency exists when the amount of this vitamin provided in the food is insufficient to meet requirements. It is accompanied by progressive exhaustion of the hepatic reserves of vitamin A. Symptoms of xerophthalmia then appear which can result in serious lesions of the cornea and possibly to destruction of the eye (keratomalacia) resulting in partial or total blindness. The keratomalacia is usually, but not necessarily, accompanied by protein calorie malnutrition of the marasmic or œdematous type. Although this association is frequent, cases of severe xerophthalmia (keratomalacia) are, however, observed in children who do not show any signs of malnutrition.

Xerophthalmia is observed most frequently in pre-school-age children between 6 months and 6 years of age. This is not surprising when it is realized that the requirements of children at this age are three times greater (per kg of body weight) than those of an adult. The infectious diseases of children, especially measles and gastro-intestinal, respiratory and urinary infections, provoke the appearance of xerophthalmia in children who have no or low hepatic reserves of vitamin A. Epidemics of xerophthalmia have been observed after bouts of diarrhoea or respiratory infections. Mortality after keratomalacia is high in children showing signs of severe malnutrition; it is higher in younger children. So far as all cases of ulceration of the cornea are concerned, it is estimated that only 30% to 35% of the children survive without adequate treatment.

Vitamin A deficiency is usually associated with respiratory or gastro-intestinal types of infectious phenomena. This can be explained in part by the fact that vitamin A helps to maintain the integrity of the respiratory and gastro-intestinal endothelium and any interference with this leads to greater vulnerability to local infections.

Vitamin A deficiency is associated with underdevelopment and is usually observed in the poorest and worst educated social strata of the populations. It was frequently observed in Europe during the pre-industrial period: it existed in Japan until 1930. It has disappeared quite recently from countries such as Korea and Singapore as a result of their accelerated socio-economic development. In many places, its existence has not been recognized owing to the fact that it affects the most socially deprived classes of the population who do not have access to health care, or because it is associated with an infectious disease which is considered to be directly responsible for blindness.

It must be emphasized that simple vitamin A deficiency increases infant morbidity and mortality.

The WHO has suggested criteria relating to the prevalence of xerophthalmia which are designed to measure its importance from the point of view of public health. These criteria are expressed in percentages of the pre-school-age population (6 months to 6 years) exposed to the risk. These criteria are the following:

Night-blindness (XN) in more than 1% of the children.

Bitot's spots (X1B) in more than 0.5% of the children.

Corneal xerosis/ulceration (1 per 10 000).

Corneal scar (XS) in more than 0.05% of the children (2 per 10 000).

In the absence of clinical symptoms of xerophthalmia, it is estimated that vitamin A deficiency constitutes a serious problem when more than 5% of the children have a plasma concentration of vitamin A lower than 10 μg/dl.

On the basis of surveys made during the course of the last ten years, it is possible to draw up a list of countries in which xerophthalmia constitutes a public health problem. In 37 countries there are sufficient indications to make it possible to assert that xerophthalmia consititutes a serious public health problem. In a certain number of cases the problem is limited to one part of the territory: such is particularly the case for certain African countries of the Sahel region, where the climatic and ecological conditions vary very greatly from the north to the south of the country and only the regions in the south are affected. In other countries where xerophthalmia is widespread, the prevalence can vary very considerably from one region to another; this is particularly the case in Indonesia where the prevalence of Bitot's spots in the population under 5 years of age varies between 0.4% and 2% (mean 1%) and that of corneal lesions from 0 to 21.2 per 10 000 (mean 6.4 per 10 000). The same variability is found in India, not only between regions but also from one year to another.

The regions most affected by xerophthalmia include southern Asia extending from Afghanistan to the Philippines, the Sahel countries of Africa and West Africa, Central America and North East Brazil.

From the point of view of numbers, an estimate based on the results of surveys carried out in India, Bangladesh and Indonesia has shown an incidence of 250 000 new cases of blindness each year in these three countries. Globally, it is estimated that every year 500 000 children suffer from active lesions of the cornea affecting one or both eyes which lead to partial or total blindness if they are not treated immediately. Approximately 2/3 of the children who become partially or totally blind die within a few weeks; thus, it is estimated that about 150 000 children survive who are partially or totally blind. Even after the initial acute phase, the mortality remains high in these children and the number who survive for 10 years is markedly reduced. That may be one of the reasons why the health authorities are so often little concerned with the problem: the evidence for its existence is lacking.

Conclusion

The 37th assembly of the World Health Organization adopted a resolution in May 1984 asking the Director General to provide Member States with all the help necessary to prevent and treat vitamin A deficiency in view of the urgency of the problem. The assembly asked him to prepare an international programme of action, in which the other agencies of the United Nations as well as non-governmental organizations would participate, and to coordinate its operation. A 10-year programme was prepared in 1985 and launched in 1986. It comprised a wide range of measures covering the long, medium and short term. At present it appears that although many countries have adopted short-term measures, few have developed long-term programmes designed to bring about permanent prevention of the deficiency. As for supplementation with high doses of vitamin A, wide-ranging programmes of prevention exist in Asia where countries such as India, Indonesia, Bangladesh, the Philippines and Nepal have undertaken to provide supplements on a national scale. In Africa, in contrast, the efforts made are still very fragmentary even though certain countries such as Malawi, Mauritania and Burkina Faso have undertaken quite extensive programmes of supplementation. It should be pointed out that in the refugee camps in Ethiopia and Sudan, supplementation with vitamin A seems at present to be carried out in an effective manner. It is perhaps of interest to report here that UNICEF/UNIPAC, Copenhagen supplied the countries in which a deficiency problem exists with an annual average of 63×10^6 capsules, each containing 200 000 IU of vitamin A, between 1985 and 1987. Certain non-governmental organizations such as the Helen Keller International, have been particularly active in the distribution of these capsules.

It is obvious that much remains to be done and perhaps the most important thing, particularly in Africa, is to arouse public opinion and the administrative and medical authorities to the importance of the problem for the health of children and the protection of their sight.

References

Amédée-Manesme O, Andersen D, Olsen JA (1988) Relation of the relative-dose-response to liver concentration of vitamin A in generally well nourished, surgical patients. Am J Clin Nutr 47:690–3

Amédée-Manesme O, Luzeau R, Carlier C, Ellrodt A (1987) Simple impression cytology method for detecting vitamin A deficiency. Lancet I:263

Amédée-Manesme O, Luzeau R, Wittpenn J, Hanck A, Sommer A (1988) Impression cytology detects subclinical vitamin A deficiency. Am J Clin Nutr 47:875–8

Barker BM, Beuder DA (1982) Vitamins in medicine. Heinemann Medical Books Ltd, London, pp211–290

Goodman DS (1974) Vitamin A transport and retinol-binding-protein metabolism. Vit Horm, 32:167–80

Goodman DS (1984) Vitamin A and retinoids in health and disease. N Engl J Med 310:1023–30

Olson JA (1978) Evaluation of vitamin A status in children. World Rev Nutr Diet 31:130

Sporn MB (1977) Vitamin A and its analogs (retinoids) in cancer prevention. Nutrition and Cancer. Wiley, New York, 119

Sommer A (1982) Nutritional blindness: xerophthalmia and keratomalacia. Oxford University Press, New York

Chapter 10

Vitamin E: the Functional Significance of Suboptimal Plasma Levels

K. F. Gey

Animal experiments suggest that aggressive oxygen species can be implicated with ageing, arterial injury by ischaemia followed by reoxygenation, arteriosclerosis, the initial states of carcinogenesis, ionizing radiation, etc. (Gey 1986). If this is applied to man these pathological processes mentioned should be inversely related to the body's multilevel defense system against free radicals. This system comprises enzymes (e.g., superoxide dismutase, catalase), endogenous non-essential antioxidants (e.g., glutathione, proteins, urate, bilirubin) and essential antioxidants (i.e., the vitamins A, C, E and carotenoids). Thus, the health hazards mentioned above should to some extent depend on the dietary supply of the latter. All essential antioxidants are liposoluble, except the water-soluble vitamin C. Vitamin E (alpha-tocopherol), the principal liposoluble antioxidant, is considered to be unique for several reasons:

1. It is an extremely potent scavanger of almost all oxygen radicals.
2. Its plasma concentration exceeds that of vitamin A and carotenoids by about 1 to 2 orders of magnitude.
3. One of its major physiological functions may consist in breaking the chain reaction of lipid peroxidation.
4. Its radical (formed by scavenging on oxygen radical) is relatively inert and can be reconverted to vitamin E by vitamin C at the interphase of membrane lipid bilayers.

Although clinical deficiency of vitamin E is rare, e.g., linked to hindrance of intestinal lipid absorption (Machlin 1984), recent epidemiological data suggest that a suboptimal plasma status of lipid-standardized vitamin E is associated with an increased risk of subsequent cancer and with an increased, age-specific mortality from ischaemic heart disease (IHD). Determination of the plasma level of vitamin E requires previous lipid-standardization, since the absolute value of vitamin E varies to some extent with the number of its carriers, i.e., mainly with the low density lipoproteins (LDL) and very low density lipoproteins (VLDL) (Gey 1986, Gey and Puska 1989).

Table 10.1. Comparison of reported and tentatively corrected plasma vitamin E in prospective studies regarding subsequent cancers. Almost all studies with statistically significant differences for mean (or median) values of cases and survivors have simultaneously indicated a significant, several-fold increased relative cancer risk at the lowest quantile (or quintile) of vitamin E levels

Study site Cancer type	Follow-up years	Plasma alpha-tocopherol (μmol/l)			
		Absolute as reported		Tentatively adjusted	
		Cancer	Survivors	Cancer	Survivors
Guernsey/Jersey, UK Breast cancer (Wald et al. 1984, 1987	10	10.9[a]	13.9	22.0	28.1
Finland Female cancers (Knekt 1988)	11	23.2[a]	24.2	31.2	33.0
Finland, low selenium All male cancers Salonen et al. 1985)	4	10.2[a]	13.0	16.0	20.4
Finland, presumably low selenium All male cancers (Knekt et al. 1988)	10	18.6[a]	19.3	24.9	26.5
Netherlands All male cancers (Kok et al. 1987)	9	16.7[a]	19.7	26.2	30.8
Washington, Md., USA Lung cancer (Menkes et al. 1986)	8	24.3[a]	27.6	28.7	32.1
Hyupertens. Progr. USA All male cancers (Willett et al. 1985)	5	26.9	29.3	29.9	32.5
Basel/Switzerland					
All male cancers (Gey et al. 1987)		33.7[a]	}	34.2	}
Gastroint. cancers (Gey et al. 1937, 1988)	7	31.2[a]	} 35.5	33.1	} 35.2
Lung cancer (Stähelin et al. 1990)		33.5[b]	}	33.7	}
All male cancers (Stähelin et al. 1989)		36.4	} 36.5	34.5	} 34.6
Lung cancer (Stähelin et al. 1989)	12	36.0	}	33.6	}
Hawaii, USA Lung cancer (Nomura et al. 1985)	10	29.7	28.6	37.1	35.7

[a] Statistically significant difference ($p<0.05$) between means or medians of cancer cases and of survivors.

[b] Statistically significant difference in Cox regression analysis (Stähelin et al. 1990).

[c] Tentative estimate of actual levels at base line by (a) linear adjustment of vitamin E from unadjusted high cholesterol levels (Knekt 1988; Knekt et al. 1988; Menkes et al. 1986) to a standard level of 220 mg/dl; (b) time-dependent decay of vitamin E by storage at −20°C of 20% (Salonen et al. 1985; Menkes et al. 1986), 36% (Kok et al. 1987), 38% (Wald et al. 1984, 1987; Knekt 1988; Knekt et al. 1988), at −70°C of 10% (Willett et al. 1985) and 20% (Nomura et al. 1985); (c) arbitrary assumption of 20% loss by freeze-thawing cycle(s) (Wald et al. 1984, 1987).

Cancer

Vitamin E can inhibit mutagenesis and carcinogenesis *in vitro* and the growth of induced tumours in many, although not all, animal models *in vivo* (Chen et al. 1988). Most prospective epidemiological studies in men have shown that lower levels of vitamin E are associated with the subsequent occurrence of cancers 4–11 years after the base-line measurement. The relative risk of subsequent cancer was 1.4- to 4.4-fold increased at the low quartile(s) or quintile(s) of vitamin E levels in comparison with high ones. This was reported for all cancers in Dutch and Finnish males (Kok et al. 1987, Knekt et al. 1988), for lung cancer in Washington County, Md. (Menkes et al. 1986), for all cancers in Finnish men with simultaneously low selenium levels (Salonen et al. 1985), for all female cancers in Finland (Knekt 1988), and possibly for breast cancer in the UK (Wald et al. 1984). However, a study on both sexes in USA (Willett et al. 1984) as well as another in males in Hawaii (Nomura et al. 1985) failed to reveal any correlation, whereas in a further study in males in Basel/Switzerland the statistical significance varied with the biostatistical model and the length of the follow-up period. For example, the 7-year follow-up for lung cancer was statistically significant in one model (Cox model of proportional hazards), but not in another (logistic regression). However, the latter showed statistical significance for the occurrence of subsequent gastro-intestinal cancers (Gey et al. 1987a,b; Stähelin et al. submitted for publication), but after 12 years no correlation could be discerned (Stähelin et al. 1989). Aside from biometric aspects, the studies differed in the length of the observation period, the storage conditions of frozen plasma, the lipid standardization, and often in the assay method. Thus only a very cautious, tentative comparison of the actual plasma status of vitamin E is possible. Nevertheless, an inverse correlation between cancer and subsequent cancer tended consistently to occur in populations for which relatively low absolute levels of vitamin E were reported, whereas populations with a higher vitamin E level showed a variable or even no association (Table 10.1). Considering follow-up periods of 4 to 10 years (which mostly revealed the greatest differences between cancer cases and survivors), a tentative estimate of corrected (conceivably actual) values suggests, in males with presumably normal selenium status, a potential protective effect of lipid-standardized vitamin E at about 30 to 35 μmol/l (13 to 19 mg/l) or more (Table 10.1).

Ischaemic Heart Disease (IHD)

There are a number of observations in favour of a role of vitamin E in IHD (Gey 1986, Gey and Puska 1989):

1. Exogenous peroxidized PUFAs damage the endothelium, proliferating smooth muscle and heart muscle; however, this damage is, at least in part, preventable by vitamin E.
2. Hypoxia with subsequent reoxygenation (reperfusion) provokes a burst of oxygen radicals which overcharge protective enzymes (superoxide dismutase,

catalase, glutathione peroxidase etc.). Arteries are severely damaged when antioxidants are exhausted.

3. Oxygen radicals crack LDL-PUFAs after loss of lipid-soluble radical scavengers (vitamin E, carotenoids). Thereby, apoprotein B is modified by hydroxyalkenals (PUFA fragments). Corresponding LDL modification by cells *in vitro* is prevented by vitamin E, also. Modified LDL is, in contrast to normal LDL, take up by monocytes/macrophages which are thus converted into arteriosclerotic lipid-laden foam cells, and this is related to the initial stages of the arteriosclerotic plaque.
4. IHD patients have increased plasma levels of thiobarbituric acid-reactive material (malondialdehyde, hydroxyalkenals) which are diminished by previous supplements of vitamin E.
5. Small amounts of modified LDL exist even in normal human plasma, but vitamin E supplements lower the level of pathological LDL.
6. Probucol, a drug with radical scavenging phenol structure, reveals in genetically hypercholesterolaemic rabbits antiarteriosclerotic properties independent of its hypocholesterolaemic potential.
7. Marginal deficiency of vitamin E causes arteriosclerotic lesions in rodents, piglets and primates. Supplements of vitamins A and/or E depress the spontaneous arteriosclerosis in older laying hens.

Cross-sectional epidemiology has accumulated further suggestive evidence that a poor plasma status of lipid-standardized vitamin E is related to a higher mortality from IHD. These comparisons were carried out in 12 European study populations, the IHD of which varied up to four-fold. The study was initiated as the International Collaborative Study on the Fatty Acid-Antioxidant Hypothesis of Arteriosclerosis, and has been continued (since 1985) as the Optional Study on Antioxidant Vitamins and PUFAs of the WHO/MONICA Project (Gey 1986, Gey et al. 1987a,c, Gey 1988, Gey and Puska 1989). The comparison of all essential antioxidants in plasma of apparently healthy males (40–49 years of age) has continuously shown the strongest association between lipid-standardized vitamin E and age-specific IHD mortality (40–59 years of age). In the latest evaluation, the monovariate analysis for vitamin E revealed an inverse correlation with $r^2=0.49$; $p<0.01$ (Gey and Puska 1989), i.e., 49% of the IHD risk may be attributable to the vitamin E level. The strength of the inverse association between vitamin E and IHD seems to be comparable to the IHD risk attributable to plasma cholesterol ($r^2=0.51$; $p<0.01$). But both risk factors may be independent, since there is also a significant inverse association between IHD mortality and absolute vitamin E levels in a subgroup of eight study populations ($r^2=0.55$; $p=0.003$), the plasma cholesterol of which does not differ significantly ($P>0.05$) and thus lacks a direct association with IHD mortality ($r^2=0.05$) in this subgroup. Plasma vitamin E does also not seem to be correlated to "classical" IHD risk factors other than lipids, i.e., not to blood pressure, smoking, body mass index, plasma glucose and selenium. However, the latest evaluation of the present study population revealed a further risk factor among the vitamins, i.e. a moderate inverse association between lipid-standardized vitamin A (retinol) and IHD mortality, i.e., with $r^2=0.33$; $p=0.07$ (Gey and Puska 1989). Since, in individuals, both lipid-soluble vitamins vary *de facto* independently (Spearman's rank coefficient $r_S=0.12$), and after lipid-standardization also independently from cholesterol (r_S<0.005), the prominent risk factors in the previous study population seem to be plasma cholesterol and the vitamins E and A. If these three risk factors are combined in a multiple regression analysis, the actual IHD mortality of all 12 study populations can be predicted almost completely (r^2=0.96; p<0.001). Therefore, IHD mortality, low by European standards (<120 IHD deaths/100 000 males), was linked to lipid-standardized plasma levels of at least 25–30 μmol/l (11–13 mg/l) of vitamin E (Gey et al. 1987a,c, Gey and Puska 1989).

Discussion of Prudent Doses

The epidemiological data available at present suggest that a suboptimal status of vitamin E may be a hitherto under-rated health hazard regarding two multifactorial diseases, i.e., cancer and IHD. Of course, a conclusive demonstration of the protective effects of an optimal status of vitamin E will in any case require population-based double-blind and compliance-controlled intervention trials with specific vitamin supplements.

Nevertheless, it may be a prudent goal to maintain a potentially protective vitamin E status, even before its preventive properties are finally proven. Other recent, but widely accepted, dietary recommendations have also been mainly based on epidemiological findings, e.g., the reduction of total lipid intakes by particular reduction of the saturated fat with the aim of getting a preponderance of monoene plus polyunsaturated lipids. The present epidemioloigcal data may suggest a potential protection at a lipid-standardized plasma alpha-tocopherol level of at least about 30–35 μmol/l (13–19 mg/l). This level also minimizes pentane exhalation, an indicator of increased lipid peroxidation due to a suboptimal status of vitamin E (Lemoyne et al. 1987). Such plasma levels may require in normal adults a minimum daily intake of about 30–60 IU vitamin E (Gey et al. 1987a, Diplock 1987). This adult requirement would be about 3–5 times that of the present RDA of 15 IU daily in the US, but still of the same order as previous RDA for the US and of reinforced recommendations of 30–45 IU daily (Get et al. 1987a).

Doses of 60–100 IU of vitamin E may be necessary to be certain of reaching plasma levels above 30 μM lipid-standardized vitamin E (Gey et al. 1987a). Moreover, an expert with the longest experience in the field suggests even a minimum of 100 IU daily for this purpose (Horwitt 1988). This may particularly be needed at a concurrently low vitamin C status and/or at increased dietary consumption of polyunsaturated fatty acids (PUFAs), as mentioned below.

In some populations studied a low vitamin C status (medians 20 to 28 μmol/l or 0.36 to 0.50 mg/dl, e.g. in Edinburgh/Scotland and Glostrup-Copenhagen/Denmark) was strongly correlated with lipid-standardized vitamin E (Spearman's rank correlation coefficient r_S=0.52) in contrast to the study populations with the highest vitamin C values (35 to 57 μmol/l, e.g. in Italy or Israel: r_S=0.15 to 0.16) (Gey and Puska 1989). Since vitamin E appears to be exclusively regenerated by vitamin C, this high correlation coefficient suggests that a poor vitamin C level limits the continuous regeneration of vitamin E, or in other words, a good vitamin E status with a poor vitamin C status may only be

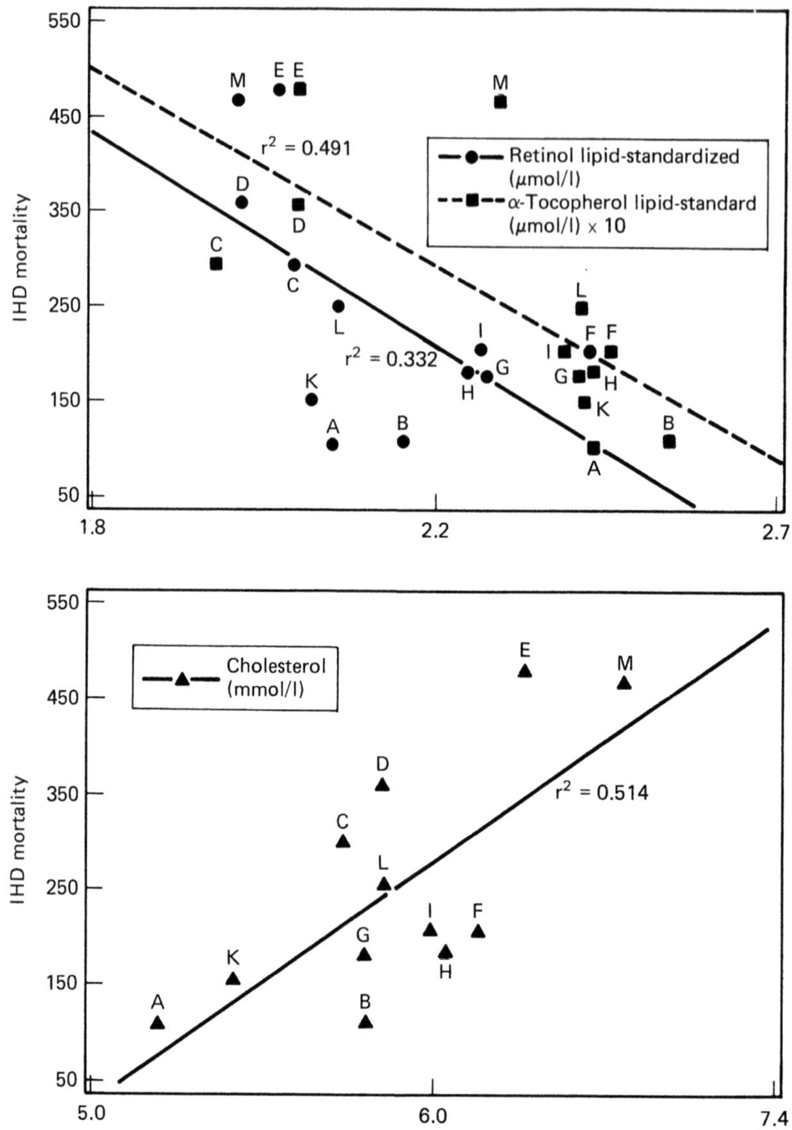

Fig. 10.1. Monovariate analysis of medians of lipid-standardized plasma vitamin E (alpha-tocopherol) and vitamin A (retinol) and of total plasma cholesterol in the whole series of study populations at present available (12 populations). The vitamins of all individuals were standardized for 220 mg cholesterol and 110 mg triglycerides prior to the calculation of medians. The ordinate gives age-specific mortality from IHD (ICD 410–414) per 100 000 males aged 50–59 years.

maintained by an increased intake of vitamin E.

Many formulas of health-oriented diets and particularly slimming diets are based on a low percentage of lipids. This may also more or less automatically reduce the intake of vegetable oils. The latter are still a major dietary source of vitamin E (Machlin 1984), although their alpha-tocopherol content has become very variable (Gey et al. 1987a, Gey 1988), due to the practices of modern food

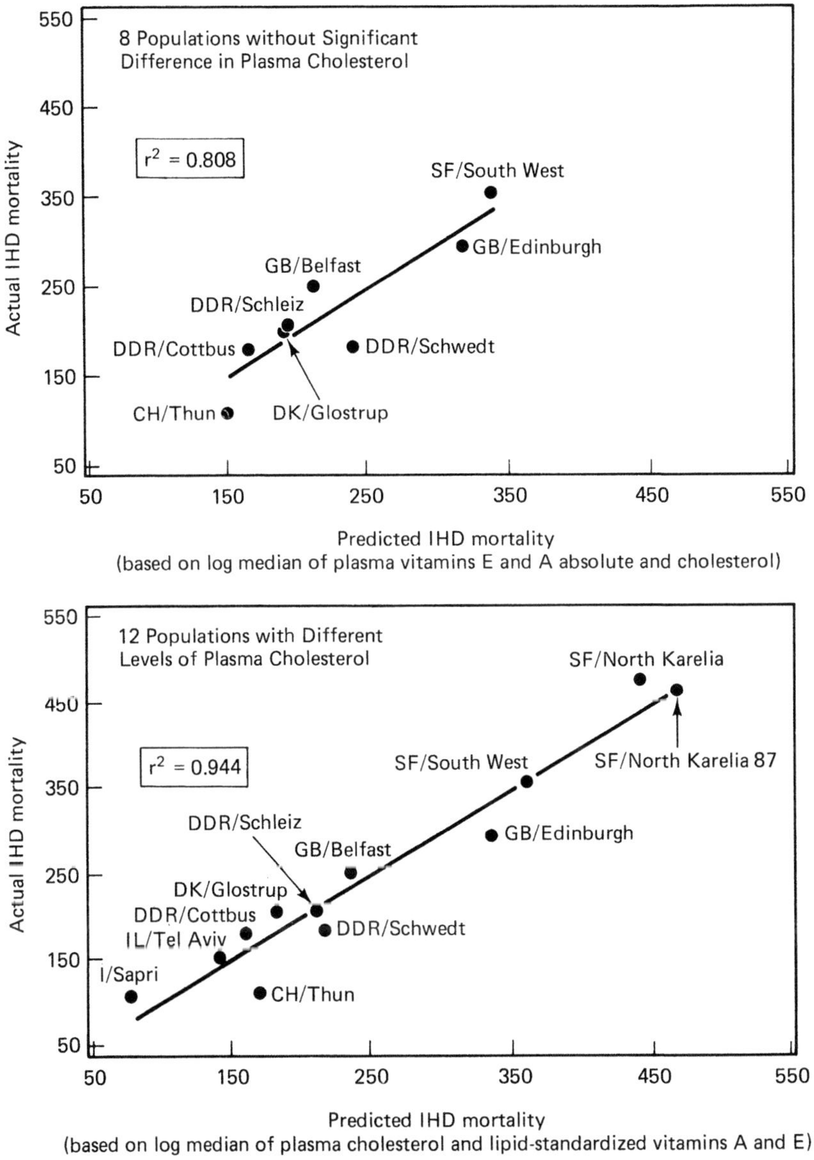

Fig. 10.2. Multivariate analysis of 8 study populations with common cholesterol levels (229–245 mg/dl, $p < 0.05$), *above*, and of all 12 study populations, *below*. When plasma cholesterol is almost identical (8 populations, *above*) its influence on IHD mortality will be very similar, and thus the latter varies primarily with the plasma vitamins E and A. In contrast, when IHD mortality is related to plasma cholesterol (12 populations, *below*) the former is significantly altered by cholesterol, too. The r^d indicates the fit between actual and predicted age-specific IHD mortality.

technology. Since retail packages of salad oils still lack a label stating the concentration of vitamin E, a consumer cannot estimate his actual dietary vitamin E intake. With these uncertainties, the consumer may not even hope substantially to improve his vitamin E status by an increased consumption of vegetable oils. Thus, the polyunsaturated fatty acids (PUFAs) of vegetable oils are generally accepted drastically to increase the vitamin E requirement (Machlin 1984). Very recently it has been attempted to reduce the risk of cardiovascular disease by using preparation of purified fish oils, which unfortunately have a low vitamin E/PUFA ratio. The relatively greater number of double bonds of PUFAs in fish oils, is however, known to require an even greater protection by vitamin E. In consequence, and because of these divergent dietary recommendations, three measures are warranted from health authorities:

1. Guidelines on how to obtain the RDA and/or a higher prudent dosage of vitamin E in cases of reduced intake of total calories and/or lipid calories.
2. A re-evaluation of the vitamin E/PUFA relationship for subjects whose dietary PUFAs of vegetable and marine origin are increased relative to saturated fat.
3. The compulsory declaration of the absolute alpha-tocopherol content and of the alpha-tocopherol/PUFA ratio in common dietary items such as vegetable oils, fish products, as well as in the ready-made food products which nowadays form a major fraction of meals in westernized societies.

An increase of vitamin E requirement through other factors, particularly growth, pregnancy and ageing, remains to be investigated in more detail. Since on the one hand classical vitamin E deficiency is characterized by severe neurological and neuromuscular defects, and since on the other hand the nervous tissue tends to take up and to retain extremely large amounts of vitamin E at the expense of all other tissues, a suboptimal vitamin status may prejudice both the normal development of the nervous system, and the protection of all its functions in later life. Interestingly, large doses of vitamin E slow significantly the progression of early Parkinson's disease (Fahn 1989).

Summary

Available epidemiological evidence suggests a decreased risk of cancer and/or ischaemic heart disease at vitamin E plasma levels of around 30–35 μmol/l (13–19 mg/l). Such levels may prudently be achieved by a regular minimum dietary intake of at least 30–60 IU, or more safely by about 100 IU vitamin E, depending on the concomitant level of dietary PUFAs and vitamin C.

References

Chen LH, Boissonneault GA, Glaurert HP (1988) Vitamin C, Vitamin E and Cancer (review). Anticancer Res 8:739–748

Diplock AT (1987) Dietary supplementation with antioxidants. Is there a case for exceeding the recommended dietary allowance? Free Radical Biol Med 3:199–201

Fahn S (1989) The endogenous toxin hypothesis of the etiology of Parkinson's disease and a pilot trial of high-dosage antioxidants in an attempt to slow the progression of the disease. Ann NY Acad Sci 570:186–196

Gey KF (1986) On the antioxidant hypothesis with regard to arteriosclerosis. Biblthca Nutr Dieta 37:53–91

Gey KF (1988) Inverse correlation of vitamin E and ischemic heart disease. Int J Vit Nutr Res 30:224–231

Gey KF, Brubacher CB, Stähelin HB (1987a) Plasma levels of antioxidant vitamins in relation to ischemic heart disease and cancer. Am J Clin Nutr 45:1368–1377

Gey KF, Brubacher GB, Stähelin JB (1987b) Cancer mortality inversely related to plasma levels of antioxidant vitamins. In: Cerutti PA, Nygaard OF, Simic MG (eds) Anticarcinogenesis and radiation protection. Plenum Press, New York London, pp 259–267

Gey KF, Stähelin HB, Puska P, Evans A (1987c) Relationship of plasma vitamin C to mortality from ischemic heart disease. Ann NY Acad Sci 498:110–123

Gey KF, Puska P (1989) Plasma vitamins E and A inversely correlated to mortality from ischemic heart disease in cross-cultural epidemiology. Ann NY Acad Sci 570:268–282

Horwitt MK (1988) Supplementation with vitamin E. Am J Clin Nutr 47: 1088–1089

Knekt P (1988) Serum vitamin E level and risk of female cancers. Int J Epidemiol 17:281–288

Knekt P, Aromaa A, Maatela J et al. (1988) Serum vitamin E and risk of cancer among Finnish men during a 10-year follow up. Am J Epidemiol 127:28–41

Kok FJ, Duijn CM van, Hofman A, Vermeeren R, Druijn AM de, Valkenburg HA (1987) Micronutrients and the risk of lung cancer. N Engl J Med 317:1416

Lemoyne M, Gossom A van, Kurian R, Ostro M, Axler J, Jeejeebhoy KN (1987) Breath pentane analysis as an index of lipid peroxidation: a functional test of vitamin E status. Am J Clin Nutr 46:267–272

Machlin LJ (1984) Vitamin E. In: Machlin LJ (ed) Handbook of vitamins. Dekker, New York Basel, pp 99–145

Menkes MS, Comstock GW, Viulleumier JP, Helsing KJ, Rider AA, Brookmeyer R (19896) Serum beta-carotene, vitamins A and E, selenium and the risk of lung cancer. N Engl J Med 315:1250–1254

Nomura AMY, Stemmermann GN, Heilbrunn LK, Salkeld RM, Vuilleumier JP (1985) Serum vitamin levels and the risk of cancer of specific sites in men of Japanese ancestry in Hawaii. Cancer Res 45:2369–2372

Salonen JT, Salonen R, Lappetelainen R et al. (1985) Risk of cancer in relation to serum concentrations of selenium and vitamins A and E: matched case-control analysis of prospective data. Br Med J 290:417–420

Stähelin HB, Eichholzer M, Lüdin E, Gey KF, Brubacher GB (1989) Cancer mortality and vitamin E status. Ann NY Acad Sci 570:391–399

Stähelin HB, Gey KF, Eichholzer M, Lüdin E, Bernasconi F, Thurneysen J, Brubacher G (1990). Plasma antioxidant vitamins and the subsequent cancer mortality in the 12 years follow-up of the Basel Prospective Study. Am J Epidemiol (submitted)

Wald NJ, Boreham J, Hayward JL, Bulbrook RD (1984) Plasma retinol, ß-carotene and vitamin E levels in relation to the future risk of breast cancer. Br J Cancer 49:321–324

Wald NJ, Thompson SG, Densem JW, Boreham J, Bailey A (1987) Serum vitamin E and subsequent risk of cancer. Br J Cancer 56:69–72

Willett WC, Polk BF, Underwood BA et al. (1984) Relation of serum vitamins A and E and carotenoids to the risk of cancer. N Engl J Med 310:430–434

Chapter 11

Selenium and Modern Lifestyles

K. H. Schmidt and W. Bayer

In relation to modern lifestyles the presence of selenium in the human environment has two aspects:

1. Selenium as a pollutant from natural and industrial sources.
2. Selenium as an essential trace element in biological systems.

Both aspects are closely interrelated owing to the fact that selenium is cycled via different pathways through soil, water, atmosphere, plants, animals and man. Quantitative data on changes over time in the selenium status of specific areas are not available. It is, however, known that areas of deficiency and of excess of selenium in the environment do occur. In addition, the availability of selenium in the soil can vary markedly and influences the selenium absorption by plants. Depending on the natural availability of selenium in the environment and on various human interventions the selenium content of food can vary over a wide range.

Whereas total selenium measured in the soil varies from 50 μg/kg to 10 000 μg/kg the percentage of water-soluble selenium available to plants is rather low (3%). In addition, the amount of selenium absorbed from the soil is not only dependent upon the available selenium pool but also upon species-specific characteristics such as, e.g., the protein content of the plants or specific accumulation and retention mechanisms. Thus the total selenium content of the soil is a poor predictor of the selenium status of the plants.

It has been shown that the selenium content of animal tissues reflects the content of this element in the feeds consumed and 5000 μg/kg feed is, at present, the accepted threshold above which toxic reactions can be expected after long-term intake. Increased retention of selenium in animal tissues under conditions where it is in low concentration has been reported, as well as reduced retention under high-intake conditions. These adaptive mechanisms are the reason for much lower fluctuations in selenium ingestion by carnivores than by herbivores.

Little information is available on the chemical forms in which selenium is present in plants. In a few studies, however, selenomethionine has been identified as the predominant selenium-containing compound in food and feedstuffs from plant origin. Much more information, however, is needed in this particular area of nutrition research.

Before selenium was recognized to be an essential nutrient for animals and man a number of selenium-responsive diseases were described in horses, cattle, and pigs. Geobotanical mapping of areas of selenium deficiency in various countries demonstrated that the geographic distribution of selenium-deficient plants correlates with a high incidence of muscular dystrophies. 100 μg selenium per kg feedstuff was identified as the lower limit of nutritional supply capable of preventing muscular dystrophy in farm animals.

Geobotanical data for selenium status are now available for a number of countries, and areas of deficiency have been identified in the USA, New Zealand, China and various parts of Europe, where the feedstuffs surveyed contained less than 20 μg/kg selenium. As an example, whole-wheat grain produced in North Dakota contains 2 ppm selenium, in New Zealand 0.1 ppm, and in the Shansi Province of China 0.005 ppm. Importation of food and feedstuffs from areas of different geobotanical selenium status can moderate this problem and markedly increase the selenium content of food supply in those countries naturally producing food low in this mineral.

A more direct way to increase the selenium content of the food supply is the addition of selenium to the feeds of farm animals. In many countries the addition of 100 μg selenium per kg feed for cattle, pigs, sheep, ducks, chickens or turkeys is accepted practice. Fortification of feedstuffs with selenium at that level is capable of preventing deficiency disorders without increasing the selenium levels in the edible tissues above that of animals fed diets naturally adequate in selenium. Again, homeostatic mechanisms play a role in this regulation.

In some countries, such as Finland, sodium selenate is added to fertilizers at a concentration of 10 mg/kg and used over the whole country. Similar effects are achieved by use of selenium containing fly ash as a soil supplement. In addition, industrial and domestic emission of selenium may contribute to the selenium content of rainwater. This is, however, frequently in the form of metallic selenium or selenium dioxide and has, therefore, a low availability to plants.

Other fertilizers such as sulphate or phosphate can decrease the selenium uptake by plants through competitive interference. Similarly, the availability of selenium is decreased by interaction with heavy metals. Muscular dystrophy due to heavy-metal-induced apparent selenium deficiency has been found in areas with extensive mining despite of adequate selenium content of the feed.

Another human activity decreasing the selenium content of food is cooking, because of the instability and volatility of selenium compounds. It has been shown that as a result of boiling vegetables up to 50% of the selenium content can be lost. Under usual cooking procedures the selenium losses of most foods are much less. Storage of foods can alter the chemical form of selenium thus influencing bioavailability. Drinking water and air do not contribute significantly to the average daily selenium intake of the general population except in seleniferous areas. Similarly, the amount of selenium inhaled with cigarette smoke is rather low.

The daily intake of selenium from dietary sources by adult human beings varies from 5000 μg/day in areas of chronic selenosis to 10 μg/day in areas of endemic cardiomyopathy (Keshan Disease). In the USA an average selenium intake of 108 μg/day was found with regional differences from 98 to 224 μg/day. Much lower dietary intakes have been found among the residents of Finland, New Zealand or West Germany.

Cereals, meat, and fish are important sources of selenium in the majority of countries. But, whereas the relative contribution of meat and fish to the total selenium intake is fairly constant, the selenium contribution of cereals shows large variations. In populations preferentially consuming wheat, corn, or rice as the natural source of dietary selenium the risk of a deficiency is much greater than in populations with a high consumption of meat and fish. Thus, differences in patterns of food consumption, individual or general, as well as differences in total food consumption due to availability or personal lifestyle profoundly affect total selenium intake.

The effect of differences in the dietary habits on the biological utilization of selenium was investigated in a number of studies with experimental diets in animals. It turned out that amino-acid based purified diets produced increased frequency and severity of selenium-responsive pancreatic atrophy in the chick (Combs et al. 1984). Feed restriction reduced the incidence of exudative diathesis in the vitamin-E-deficient chick (Zhou and Combs 1984).

A low protein diet increases tissue levels of selenium if the total selenium intake is low (Zhou et al. 1983). High sulphate ingestion resulted in a significantly increased urinary excretion of selenium in the rat (Ganther and Baumann 1962). With a high-fat diet enteric absorption of selenium was reduced (Mutanen and Mykkanen 1984). Vitamin B6-deficient diets significantly reduce the utilization of selenomethionine in the liver of rats (Yasumoto et al. 1979). Similarly, riboflavin deficiency was shown to reduce selenium contents in the livers of pigs (Brady et al. 1979). Vitamin E deficiency reduces enteric absorption as well as metabolic utilization of selenium (Lane et al. 1979).

A synergistic effect of ascorbate and selenium was observed in the protection of vitamin E-deficient chicks from exudative diathesis (Combs and Scott 1974). Similar effects were seen with other antioxidants such as ethoxquin, DPPD, BHT, or DHA (Bieri 1964).

In contrast heavy metals, such as mercury, significantly increased the retention of selenium.

Similar studies in humans are not available at the present time but similar interactions between diet and selenium status can be expected. In addition to the dietary habits the nutritional selenium status can also change in relation to different life cycle phases. A typical example is the lower selenium level in the plasma of pregnant women as compared to non-pregnant.

Other studies have shown that the selenium status is affected by specific drugs such as for example oral contraceptives or aurothiomalate but also by intoxicants such as ethanol. In addition, physical stresses such as ionizing radiation can affect the selenium status of man.

In comparative studies it was found that the selenium levels in blood as well as glutathione peroxidase activities of elderly people were significantly lower than those in young adults. From these studies it could not be differentiated whether the low selenium levels were caused by low intake or by the ageing process itself. However, assessment of 7-day pooled dietary composites of elderly people in Sweden revealed an average daily selenium intake of 30 μg, which is well below the lower limit of the recommended daily intake.

The selenium supply to infants and small children is critical because of their high metabolic turnover and their nutritional reliance on milk, the selenium content of which can vary significantly. Depending on the geographical area, differences in the selenium content of mother's milk between 2 μg/l and 300 μg/l were found.

Nutritional deficiences in the selenium supply are associated with a number of health risks among which Keshan disease shows the clearest association with a low selenium status. In addition to heart damage, endemic osteoarthropathy is found in low selenium areas as well as a higher incidence of myocardial ischaemia and atherosclerosis. Furthermore, studies in experimental animals provide evidence for an increased cancer risk in animals fed on a low selenium diet compared to animals whose feed is supplemented with selenium. Three epidemiological studies showed a consistent inverse association between serum-selenium level and the risk of cancer, although this association was only significant in smokers and in males. From all studies carried out it is very likely that other factors are involved in the aetiopathogenesis of the diseases in addition to the low selenium status. Much more research is needed in order to clarify the role of these factors in relation to selenium.

References

Bieri JG (1964) Synergistic effects between antioxidants and selenium or vitamin E. Biochem Pharmacol 13:1465–1469

Brady PS, Brady LJ, Parsons MJ, Ullrey DE, Miller ER (1979) Effects of riboflavin deficiency on growth and glutathione peroxidase system enzymes on the baby pig. J Nutr 109: 1615–1622

Combs GF, Scott ML (1974) Antioxidant effects on selenium and vitamin E function in the chick. J Nutr 104:1297–1303

Combs GF, Liu CH, Lu ZH, Su Q (1984) Uncomplicated selenium deficiency produced in chicks fed a corn-soy based diet. J Nutr 114:964–976

Ganther HE, Baumann CA (1962) Selenium metabolism. II. Modifying effects of sulphate. J Nutr 77:408–413

Lane HW, Shirley RL, Cerda JJ (1979) Glutathione peroxidase activity in intestinal and liver tissues of rats fed various levels of selenium sulphur and alpha-tocopherol. J Nutr 109:444–452

Mutanen ML, Mykkanen HM (1984) Effect of dietary fat on plasma glutathione peroxidase levels and intestinal absorption of ^{75}Se-labelled sodium selenite in chicks. J Nutr 114:829–834

Yasumoto K, Iwami K, Yoshida M (1979) Vitamin B6 dependence of seleno-methionine and selenite utilization for glutathione peroxidase in the rat. J Nutr 109:760–766

Zhou YP, Combs GF (1984) Effects of dietary protein level and level of food intake on the apparent bioavailability of selenium for the chick. Poultry Sci 63:294–303

Zhou R, Sun S, Zhai F, Man R, Guo S, Wang H, Yang G (1983) Effect of dietary protein level on the selenium contents and glutathion peroxidase activities in blood and tissues of rats. Yingyang Zuebao 5:145–151

Chapter 12

Calcium: The Functional Significance of Trends in Consumption

A. Prentice

Modern Trends in Calcium Intakes

The modern trend towards decreased energy intakes in industrialized countries has been accompanied by a steady decline in calcium consumption. Data from the British National Food Survey (MAFF 1970–1985) show that in the 15 years since 1970 calcium intakes in Britain have fallen by 20% (Fig. 12.1). In common

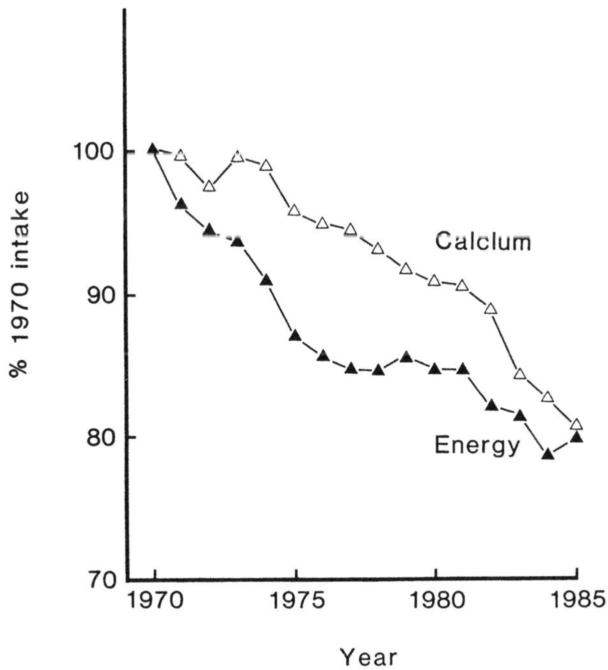

Fig. 12.1. Decline in British calcium and energy intakes between 1970 and 1985. Data from the British National Food Survey (MAFF 1970–1985)

with many other European countries, Australia, New Zealand and the USA about two-thirds of the calcium in the British diet is provided by milk and milk products (yoghurt, cheese, etc.) (FAO 1977, MAFF 1985) and falling consumption of liquid milk accounts for a large proportion of the decreased calcium intakes in recent years (Fig. 12.2, Table 12.1). The decrease in milk consumption has been influenced by the increased availability of soft drinks and fruit juices, by increases in the price of milk and by heightened interest in dietary aspects of obesity and coronary heart disease (Buss 1988). In contrast to their contribution to calcium intake, milk and milk products are only minor contributors to energy provision and the decline in energy intakes is largely due to changes in sugar, meat and bread consumption (Fig. 12.2, Table 12.1). The fall in energy intakes in Britain preceded the decline in milk consumption by about 5 years and consequently the calcium density of the British diet (mg/1000 kcal) rose during the early nineteen seventies and still remains slightly elevated compared with 1970 (Fig. 12.3).

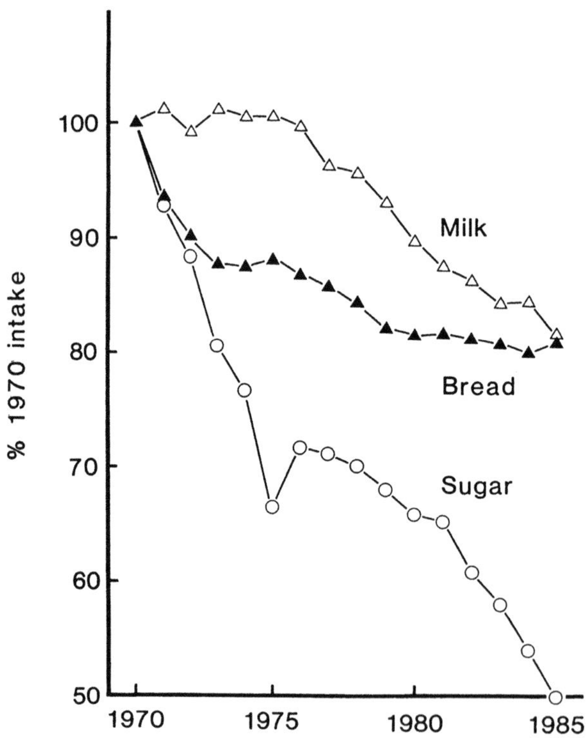

Fig. 12.2. Changes in consumption of major calcium and energy containing foods in Britain between 1970 and 1985. Data from the British National Food Survey (MAFF 1970–1985)

Table 12.1. Contribution of different foods to the decline in calcium and energy intakes in Britain between 1970 and 1984. Data from the British National Food Survey

	Calcium (Δ%[a])	Energy (Δ%)
Milk and milk products	71.6	8.4
Cereals including bread	22.4	28.7
Fruit, vegetables, potatoes	4.4	1.9
Eggs	3.8	2.8
Fish	1.1	-0.2
Fats	0	12.4
Sugar and preserves	-0.5	24.7
Meat	-3.3	20.2
Others	0.5	1.1

[a] Δ%, percentage contribution to decrement in calcium consumption between 1970 and 1984. A negative value indicates an increased intake of a particular foodstuff.

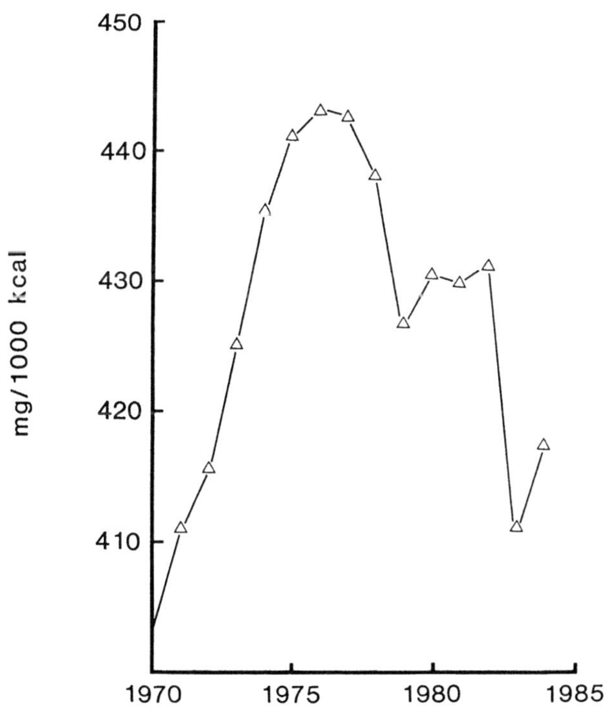

Fig. 12.3. Changes in the calcium density of the British Diet between 1970 and 1985. Data from the British National Food Survey (MAFF 1970–1985)

International Perspective on Calcium Intakes

The decline in calcium intakes that has occurred in industrialized countries is relatively small when compared with international differences in calcium consumption. In particular Western Europe, USA and Australasia have habitual calcium intakes that are two to three times greater than in Africa and the Far East (Table 12.2). These differences are primarily related to the level of milk consumption. Milk supply varies over a threefold range throughout the world and has changed little over the past 30 years (Table 12.3). In non-milk-drinking populations the main sources of calcium are leaves, cereals, legumes, fish and, in certain areas, drinking water (Walker 1972). Not only is calcium provision from these sources low but calcium bioavailability may be poor due to high phytate, oxalate and fibre contents and the absence of absorption-promoting factors, such as lactose, in the diet (Allen 1982). Immigrants to industrialized countries from non-dairying societies often retain their dietary traditions and calcium intakes remain low in spite of high availability of milk and milk products (Kim et al. 1984; M'Buyamba-Kabangu et al. 1987).

The differences between milk-drinking and non-milk-drinking societies are further emphasized when comparing pre-menopausal women. In many parts of the developing world, unlike industrialized countries, it is the norm for women of reproductive age to be either pregnant or lactating. Extra or preferential food consumption to cover the increased calcium required for fetal growth and milk synthesis is rare. Recent studies, for example, comparing calcium intakes during lactation in rural West African and British women have observed a threefold difference in calcium intake between the two (500 mg/day and 1500 mg/day) (Black et al. 1986; MA Laskey and A Prentice unpublished).

Table 12.2. International calcium intakes and age-adjusted hip-fracture incidence

	Calcium consumption[a] (mg/caput/day)	Hip-fracture incidence[a] per 100 000 (women)
Finland	1430	50
Sweden	1180	87
United Kingdom	1000	63
United States	970	102
New Zealand	970	97
Jerusalem	970	70
Holland	810	51
Singapore	660	15
Hong Kong	430	32
South African Bantu	175–475[b]	5

[a] Data originally from FAO 1977 and Gallagher et al. 1980 summarized by Hegsted 1986 and the Royal College of Physicians 1989.

[b] Range estimated by Walker (1972).

Table 12.3. World supply of energy from milk. Data from the Fifth World Food Survey 1985

	Energy supplied from milk (kcal/caput/day)		
	1961–1963	1969–1971	1979–1981
World	117	116	116
Developed market economies[a]	281	280	287
Eastern Europe and USSR	273	310	291
Latin America	119	130	144
Near East	85	82	105
Far East	52	46	54
Africa	38	41	45
ACPE[b]	9	8	12

[a] North America, Western Europe, Oceania.
[b] Asian centrally-planned economies: China, Kampuchea, Korea, Mongolia, Vietnam.

Calcium deficiency

The human body contains approximately 25g of calcium at birth and this increases to about 1300 g at maturity. Over 99% of body calcium is contained within bone and teeth where it provides strength and structure. Additionally, calcium is present in all living tissues where it acts as a regulator of metabolism and is essential for such body processes as blood-clotting, neurotransmission and muscle contraction. During pregnancy and lactation substantial quantities of calcium are required for fetal growth and breast-milk production.

As all the calcium required for maintenance, growth and reproduction must be furnished from dietary sources inevitably an insufficient dietary calcium supply must result in perturbations of calcium metabolism, bone imperfections and impaired reproductive performance. Calcium deficiency secondary to defective calcium absorption, hormonal imbalances and excessive calcium excretion such as occur in a variety of human disease states (e.g., renal failure, gastro-intestinal resection, parathyroid and thyroid dysfunction, vitamin D deficiency) lead to bone abnormalities. Similarly in very low birthweight babies inappropriate calcium supply and problems of calcium handling are associated with osteopenia and reduced linear growth (Brooke and Lucas 1985). However, in otherwise healthy individuals, it is difficult to show any immediate detrimental effect of low calcium intakes and the involvement of calcium deficiency in the aetiology of chronic human disease is poorly understood.

Osteoporosis

Osteoporosis is a progressive loss of bone tissue which is accompanied by an increased risk of bone fracture after minimal trauma. Bone loss is a feature of middle and old age in both sexes but the rate of loss is particularly high in women following the menopause. Post-menopausal osteoporosis occurs predominantly in trabecular bone and is associated with spine and wrist fractures whereas hip (neck of femur) fractures in the elderly are associated with cortical and trabecular bone loss (Kanis 1984; Dequeker 1988). Although the strength of a bone is related to its mineral content (Dequeker 1988), an individual's bone density is only one factor which may influence fracture risk and there is substantial overlap in measured bone mineral contents between people who fracture their bones and those who do not (Dequeker 1988). Many factors, such as oestrogen and vitamin D status, lack of physical activity, smoking and alcohol consumption, have been implicated in the aetiology of hip fracture (Royal College of Physicians 1989). The role of current, recent and lifelong calcium intakes in skeletal mineralization, bone loss and fracture incidence has long been assumed to be important but evidence in support of this concept is confused and contradictory. Indeed when viewed on an international basis the paradoxical situation exists where the lowest incidence of hip and other fractures occurs in those countries with the lowest calcium intakes (Table 12.2; Walker 1972; Heaney et al 1982; Royal College of Physicians 1989).

The calcium intakes of osteoporosis patients tend to be lower than controls both at the time of fracture and in recent years (Heaney et al. 1982; Cooper et al. 1988; Holbrook et al 1988; Lau et al. 1988). Most studies on healthy post- or peri-menopausal women, however, have found either no correlation between current calcium consumption and rate of bone loss or bone mineral content at a variety of skeletal sites (Smith and Frame 1965; Freudenheim et al. 1986; Riggs et al. 1987; Angus et al. 1988; Stevenson et al. 1988) or only weak relationships (Yano et al. 1985).

It has been suggested and generally accepted that the mineral content of the skeleton at the end of the growth and consolidation phases (30–40 years) is a primary determinant of bone density and fracture risk in later life (Newton-John and Morgan 1968; Dequeker 1988; Royal College of Physicians 1989). The influence of calcium nutrition in childhood, adolescence and early adult life on peak bone mass and its relationship to fracture risk is not known. Studies in the USA monitoring bone density in post-menopausal women and in men and women aged 15–90 years have demonstrated positive correlations with remembered calcium intakes in childhood or over a lifetime (Hurxthal and Vose 1969; Sandler et al. 1985; Cauley et al. 1988). Pre-menopausal bone density in North American women has been related to current calcium intakes (Kanders et al. 1988; Picard et al. 1988). Prolonged breast-feeding, which involves the loss of 200–300 mg-day into milk, has been associated with reduced bone densities at the age of peak bone mass (Wardlaw and Pike 1986) and progressive bone loss during lactation has been recorded in American teenage mothers on low calcium intakes (Chan et al. 1987). Recent studies from The Gambia, however, have shown that rural African mothers, who in the course of a reproductive life are likely to have over 10 children, each of whom will be breast-fed for at least 2 years, show the same pattern of bone mineral increase up to a peak in the fourth decade with little difference in absolute peak bone mineral content than American and British women (J Shaw, MA Laskey and A Prentice unpublished). In one much-quoted study from Yugoslavia (Matkovic et al. 1979) metacarpal cortical thicknesses in middle life were greater in a community consuming a calcium-rich diet compared with an outwardly similar area with a calcium-poor diet. The incidence of hip fracture in the elderly was less in the high calcium area but the incidence of wrist fracture was the same. As a number of authors have pointed out (e.g., Stevenson et al. 1988; Kanis and Passmore 1989a), energy intakes, and therefore possibly energy expenditures, also differed between the communities and this may have contributed to the observed differences in bone indices and fracture incidence. Indeed, in most studies, calcium consumption is likely to act as a proxy for other differences in general nutrition and lifestyle and consequently relationships between calcium intake and bone density or fracture risk cannot be interpreted as causal.

Calcium supplementation of fracture patients and healthy post-menopausal women in industrialized countries appears to have little effect on the rate of bone loss from sites containing high proportions of trabecular bone but small reductions in cortical bone loss have been recorded in some studies (Table 12.4). A number of treatment regimes cause reductions in the rate of post-menopausal bone loss but it is at present unclear whether this ultimately translates into decreased fracture incidence. It has been argued that some of these treatments may be deleterious as they may alter bone composition, the processes of microfracture repair or cause a re-distribution of mineral between skeletal sites (Kanis 1984; Kanis and Passmore 1989b).

Growth

The influence of low calcium intakes on growth and bone development in children is unknown although impaired calcium metabolism in premature babies affects both bone composition and linear growth (Brooke and Lucas 1985). In many developing countries, where calcium intakes are low, growth is often retarded due to a combination of high infection rates and poor general nutrition. Calcium supplementation studies in such children have generally failed to detect any change in growth rate (Walker 1954; Walker 1972; Kanis and Passmore 1989b). Single photon absorptiometry measurements on the mid-radius of rural Gambian children consuming 200 mg calcium/day have shown a retardation in bone development compared with age-matched British children but this difference largely disappears when the children are compared on a body size basis (MA Laskey, J Shaw and A Prentice unpublished). In The Gambia and other developing countries the time to maturity is longer than in industrialized countries and the slower rates of growth and bone development do not appear to grossly prejudice adult height or bone composition (Walker 1954, 1972). Very low calcium intakes have been implicated in the aetiology of a vitamin-D resistant rickets which has been observed in a number of tropical countries where sunlight exposure is unrestricted (Pettifor et al. 1979; Marie et al. 1982) and in calcium-deprived young baboons (Pettifor et al. 1984). The threshold of calcium intake below which bone abnormalities may occur is not known but the scarcity of reported cases of calcium-responsive rickets suggests that it is lower than the calcium content of the majority of diets throughout the world.

Hypertension

Cross-sectional epidemiological evidence from the USA and Europe have shown an association between low current calcium intakes and hypertension (Karanja and McCarron 1984). Although this may partially reflect a conscious avoidance

Table 12.4. Effect of calcium supplements alone on rate of bone loss in postmenopausal women.

Country	Subjects[a]		Duration (years)	Calcium intakes mg/day		Skeletal site[b]	Measurement[c]	Effect[d]	Reference
	Supplement	Control		Initial	Supplement				
Denmark	14 PM	11 PM	2	nd	+2000	Mid radius C	SPA	+	Riis et al. 1987
						Distal radius T	SPA	0	
						Spine T	DPA	0	
						Total body C	DPA	+	
Sweden	25 Os	25 Os	3	nd	+1000	Spine T	DPA	0	Hansson and Roos 1987
Australia	136 PM	74 PM	1.5	711	+890	Forearm C	SPA	+	Polley et al. 1987
USA	34 PM	33 PM	4	723	+1273	1/3 ulna C	SPA	0	Freudenheim et al. 1986
						1/3 radius C	SPA	(+)	
						Mid humerus C	SPA	(+)	
USA	30 PM	13 PM	2	nd	+1500	Spine T	QCT	0	Gordon and Genant 1985
USA	13 PM	9 PM	1	679	+792	Distal forearm T	SPA	0	Recker and Heaney 1985
						Metacarpal C	RG	0	
Denmark	103 PM	–	2	430, 880, 1150	+500	Distal forearm T	SPA	0	Nilas et al. 1984
USA	22 PM	20 PM	2	503	+1040	Distal radius T	SPA	0	Recker et al. 1977
						Metacarpal C	RG	+	
UK	24 PM	18 PM	2	nd	+800	Distal ulna T	SPA	+	Horsman et al. 1977
						Distal radius T	SPA	0	
						Metacarpal C	RG	0	

[a] PM, Healthy postmenopausal women; Os, osteoporosis patients.

[b] T, High proportion of trabecular bone; C, high proportion of cortical bone.

[c] SPA, Single photon absorptiometry; DPA, dual photon absorptiometry; QCT, quantitative computer tomography; RG, radiogrammetry.

[d] +, Significant reduction in rate of bone loss; (+), no clear effect of supplementation but a significant correlation between calcium intake and rate of bone loss in control group or in supplemented and control groups combined; 0, no significant reduction in rate of bone loss.

of milk and milk products after the diagnosis of hypertension, there is limited evidence from longitudinal supplementation studies that increasing calcium intakes may cause a decrease in blood pressure in a proportion of hypertensives (Karanja and McCarron 1984; Johnson et al. 1985). As with osteoporosis, however, on an international basis age-related hypertension is a feature of populations with the highest habitual calcium intakes. The incidence of pregnancy-induced hypertension, however, is high in countries with low calcium intakes even when differences in ante-natal care, age and parity are treated as confounders and supplementation of pregnant mothers with large doses of calcium (2 g/day) may prevent the rise in blood pressure during the third trimester (Karanja and McCarron 1984; Belizan et al. 1988). The evidence that calcium is important in the aetiology and treatment of hypertension can at present only be regarded as speculative.

Colorectal Cancer

Colorectal cancer is a major problem in populations consuming Western-style diets and has been associated with high intakes of dietary fat. A hypothesis has been advanced that high intakes of calcium may attenuate the carcinogenic effects of fat in the large bowel (Newmark et al. 1984). Evidence in support of this has come from animal studies but as yet results of limited human studies are conflicting (Heilbrun et al. 1986; Newmark and Wargovich 1986).

Adaptation or Altered Requirements?

The large disparity in calcium intakes which occurs throughout the world with little apparent effect on health suggests either that man can adapt to varying levels of calcium in the diet or that certain facets of life in different populations may alter the calcium requirement.

Balance studies indicate that a high proportion of subjects are in negative calcium balance at low calcium intakes and that in general zero calcium balance occurs between 400 and 800 mg/day (Irwin and Keinholtz 1973; Marshall et al. 1976; Spencer et al. 1982, 1984; Heaney et al. 1982; Nordin and Marshall 1988). The between-individual variation, however, is enormous and some subjects are in positive balance at very low calcium intakes while others are in negative balance at intakes above 1500 mg/day (Irwin and Keinholz 1973; Marshall et al. 1976; Spencer et al. 1984). Calcium retentions, and hence apparent calcium requirements, increase as calcium consumption increases reaching a plateau at high intakes (Leitch and Aitken 1959; Irwin and Keinholz 1973; Spencer et al. 1984; Kanis and Passmore 1989a) and, curiously, subjects tested at a variety of intakes tend to have higher retentions at a specific intake when having recently changed from a lower calcium intake than after moving from a higher intake (Leitch and Aitken 1959). These findings strongly suggest that calcium balance is a function of previous diet history (Kanis and Passmore 1989a; Jeejebhoy 1986). It is noteworthy that the majority of balance studies have been conducted on subjects from Europe and USA who were likely to have been habituated over a lifetime to relatively high calcium intakes and who were introduced to lower or higher intakes only shortly before the investigation.

A balance study conducted in a country with low habitual calcium intakes was performed in Peru in which zero balance was recorded at about 200 mg/day (Hegsted et al. 1952). Although this study has often been discounted on technical grounds (Nordin and Marshall 1988), it does lend support to the concept that previous diet history may determine calcium balance. Further evidence is provided by a longitudinal study of Norwegian men who were investigated after their calcium intakes had been substantially reduced (Malm 1958; Kanis and Passmore 1989a). The subjects were in negative calcium balance for several months but the calcium loss diminished with time and eventually equilibrium was re-established in the majority of subjects (88%). Other authors, however, state that they have failed to detect adaptation in subjects studied for prolonged periods (Heaney et al. 1982; Spencer et al. 1984), suggesting that the ability to adapt may vary from one individual to another or during an individual's lifetime. It has been argued, however, that if adaptation is mediated by changes in bone remodelling it could take several years to re-establish equilibrium because of the slow turnover rate of cortical bone (Kanis 1984; Kanis and Passmore 1989a,b).

It is possible that the calcium requirement or the ability to adapt to changes in calcium intake may vary considerably between populations or groups of individuals within populations. This might reflect genetic disposition or differences in lifestyle, particularly diet composition and physical activity, which influence calcium absorption and obligatory calcium excretion.

Absorption

Current thought is that calcium is absorbed from the gastro-intestinal tract by two mechanisms, one passive and one active (Bronner 1988). The passive transport mechanism is paracellular, operates throughout the gut and is not under hormonal or vitamin D control. The percentage of luminal calcium transported remains constant at all levels of calcium concentration and the absolute amount of calcium transported, therefore, increases as the luminal calcium content rises. The active mechanism is transcellular, occurs predominantly in the small intestine, and involves the intracellular protein CaBP (calcium binding protein). The synthesis and activity of CaBP is under hormonal and vitamin D control. This process is saturated at intraluminal calcium concentrations above a certain limit and its contribution to absorption diminishes at high calcium concentrations. Calcium intakes and changes in calcium requirement may influence the efficiency of calcium absorption. In rats low calcium intakes during growth induce higher levels of intracellular CaBP and enhanced active transport (Pansu et al. 1981, 1983; Bronner 1988) while pregnancy increases both active and passive absorption (Allen 1982). Increases in calcium absorption have been recorded in human pregnancy (Heaney and Skillman 1971) but little is known about the potential for CaBP induction in the human intestine and whether high habitual calcium intakes may blunt any CaBP response to changes in intake or requirement.

Other components of the diet can influence calcium absorption either by reducing the amount of luminal calcium available for absorption e.g., phosphate, phytate, fibre and oxalate or by altering the efficiency of absorption across the gut (Allen 1982; Bronner 1988). Substances which increase absorption efficiency include simple sugars and amino acids, those which decrease it include theophylline and alcohol. In many cases substances which alter calcium availability are associated with parallel changes in urinary excretion (Peacock 1988) and calcium retention remains essentially unchanged. It is now thought unlikely that in healthy individuals high intakes of phosphate, phytate and oxalate, major components of vegetarian diets, are likely to substantially influence calcium retention (Allen 1982; Bronner 1988).

Excretion

Two major pathways exist for calcium excretion, via the urine and via the gut (Peacock 1988). Endogenous calcium is excreted into the gut in digestive juices, some of which is re-absorbed and re-utilized and some of which appears in the faeces. Little is known about the effect of calcium intakes on the incorporation of calcium into digestive juices but any change in absorption efficiency will alter the amount of endogenous calcium excreted in the faeces.

Many dietary factors and common self-medications influence the quantity of calcium excreted on a daily basis into urine. Dietary substances which increase calcium losses include calcium itself, sodium, protein, caffeine and antacids while boron and phosphorus decrease calcium excretion (Spencer and Kramer 1986; Nielsen et al. 1987; Massey and Hollingbery 1988; Nordin and Marshall 1988; Peacock 1988). The relationship between urinary calcium output and dietary calcium intake can be modified by substances which alter calcium availability or the efficiency of absorption (see above). The effects of dietary factors which increase obligatory calcium losses can be sizeable and calcium excretion in populations consuming low levels of salt, protein and coffee may well be substantially lower than in Western countries. It has been stated that urinary calcium is almost more dependent on salt and protein intake than it is on calcium intake (Nordin and Marshall 1988). Periods of immobility are associated with increased loss of calcium from the body via the urine, whereas physical activity leads to enhanced calcium retention (Irwin and Keinholz 1973; Heaney et al. 1982). It is, therefore, possible that the composition of the diet in industrialized countries and other facets of modern lifestyles, including the trend towards highly sedentary lives, may lead to greater obligatory losses of calcium in the urine and hence to a higher calcium requirement.

Calcium RDAs and the Significance of Decreased Calcium Intakes

The lack of any firm evidence linking low calcium intakes to human disease, the differences in calcium supply around the world, the potential ability to adapt to changes in calcium intakes and the possible impact of race and lifestyle on calcium requirements make the evaluation of recommended daily amounts (RDAs) for calcium extremely difficult. At present there is no international consensus on requirements and RDA values recommended by different advisory bodies vary over a two- to threefold range (Table 12.5; Truswell et al. 1983). The RDA for lactating women, for example, is 600 mg calcium/day in Indonesia, 1200 mg/day in the UK and 2000 mg/day in Poland. Considerable disparities are found even between neighbouring countries in Western Europe.

The British National Food Survey calculated the average calcium consumption in Britain to be 194% of the British RDA value in 1970 (MAFF 1970). By 1985 this had decreased to 163% (MAFF 1985). However, if the British consumptions had been expressed relative to the RDA value of the Netherlands, France or Germany, the 1985 intake would be about 100% RDA and further decreases would lead to values below the recommended amount.

It can be seen from the foregoing discussion that there are no accepted criteria on which to assess the functional significance of recent trends towards lower calcium intakes in industrialized countries. Before such an evaluation can be made much more information is required on the relationships between intake, calcium metabolism and health and on the capacity of individuals to adapt to changes in calcium intake. Future researches should involve all stages of life, particularly times of increased requirement such as growth, pregnancy and lactation, and valuable insight will be gained from cross-country studies comparing populations with different lifestyles and calcium nutrition.

Table 12.5. Recommended dietary intakes of calcium. Data from the Report of Committee 1/15 of the International Union of Nutritional Sciences 1982 (Truswell et al. 1983)

	Infants (6–12 mo)	Children	Puberty[a]	Men	Women	Pregnancy[b]	Lactation
China 1981	600	700	1200	600	600	1500	2000
FAO/WHO 1974[c]	550	450	650	450	450	1100	1100
France 1981	600	700	1000	800	800	1000	1200
West Germany 1975	500	700	1000	800	500	1200	1200
India 1981[c]	550	450	650	450	450	1000	1000
Indonesia 1980	600	500	700	500	500	600	600
Italy 1978	600	500	700	500	500	1200	1200
Japan 1979	400	400	900	600	600	1000	1100
Netherlands 1978	100/kg	800	1200	800	800	1300	1500
Poland 1969	–	1000	1400	800	800	1400	2000
Thailand 1970	500	400	700	500	400	1000	1200
United Kingdom 1979	600	600	700	500	500	1200	1200
USA 1980	540	800	800	800	800	1200	1200
USSR[d]	1000	1000	1500	800	800	1500	1900
Western Pacific 1972	400	400	600	400	400	600	600

[a] Peak recommendation for boys.

[b] Last trimester of pregnancy.

[c] Mid-value of range quoted.

[d] Date not given in original source.

Acknowledgement. I would like to thank Miss A.A. Paul for the analysis of the British National Food Survey data.

References

Allen LH (1982) Calcium bioavailability and absorption: a review. Am J Clin Nutr 35:783–808
Angus RM, Sambrook PN, Pocock NA, Eisman JA (1988) Dietary intake and bone mineral density. Bone and Mineral 4:265–277
Belizan JM, Villar J, Repke J (1988) The relationship between calcium intake and pregnancy-induced hypertension: up-to-date evidence. Am J Obstet Gynecol 158:898–902
Black AE, Wiles SJ, Paul AA (1986) The nutrient intakes of pregnant and lactating mothers of good socio-economic status in Cambridge: some implications for recommended daily allowances of minor nutrients. Br J Nutr 56:59–72
Brooke O, Lucas A (1985) Metabolic bone disease in preterm infants. Arch Dis Child 60:682–685
Bronner F (1988) Gastrointestinal absorption of calcium. In: Nordin BEC (ed) Calcium in human biology. Springer-Verlag, Berlin, pp 93–118
Buss DH (1988) Is the British diet improving? Proc Nutr Soc 47:295–306
Cauley JA, Gutai JP, Kuller LH, LeDonne D, Sandler RB, Sashin D, Powell JG (1988) Endogenous Oestrogen levels and calcium intakes in postmenopausal women. Relationships with cortical bone measures JAMA 260:3150–3155
Chan GM, McMurray M, Westover K, Engelbert-Fenton K, Thomas MR (1987) Effects of increased dietary calcium intake upon the calcium and bone mineral status of lactating adolescent and adult women. Am J Clin Nutr 46:319–323
Cooper C, Barker DJP, Wickham C (1988) Physical activity, muscle strength and calcium intake in fracture of the proximal femur in Britain. Br J Med 297:1443–1446
Dequeker J (1988) Calcified tissues: structure-function relationships. In: Nordin BEC (ed) Calcium in human biology. Springer-Verlag, Berlin, pp 209–230
FAO (1977) The fourth FAO world food survey. FAO, Rome
FAO (1987) The fifth FAO world food survey. FAO, Rome
Freudenheim JL, Johnson NE, Smith EL (1986) Relationships between usual nutrient intake and bone-mineral content of women 35–65 years of age: longitudinal and cross-sectional analysis. Am J Clin Nutr 44:863–876
Gallagher JC, Melton LS, Riggs BL, Bergstrath E (1980) Epidemiology of fractures of the proximal femur in Rochester, Minnesota. Clin Orthop Relat Res 150:163–171
Gordon GS, Genant HK (1985) The ageing skeleton. Clin Geriat Med 1:95–118
Hansson T, Roos B (1987) The effect of fluoride and calcium on spinal bone mineral content: a controlled, prospective (3 years) study. Calcif Tissue Int 40:315–317
Heaney RP, Skillman TG (1971) Calcium metabolism in normal human pregnancy. J Clin Endocrinol-Metab 33:661–670
Heaney RP, Gallagher JC, Johnston CC, Neer R, Parfitt AM, Whedon GD (1982) Calcium nutrition and bone health in the elderly. Am J Clin Nutr 36:986–1013
Hegsted DM, Moscoso I, Collazos C (1952) A study of the minimum calcium requirements of adult men. J Nutr 46:181–201
Hegsted DM (1986) Calcium and osteoporosis. J Nutr 116:2316–2319
Heilbrun LK, Hankin JH, Nomura AMY, Stemmermann GN (1986) Colon cancer and dietary fat, phosphorus and calcium in Hawaiian-Japanese men. Am J Clin Nutr 43:306–309
Holbrook TL, Barrett-Connor E, Wingard DL (1988) Dietary calcium and risk of hip fracture: 14 year prospective population study. Lancet ii:1046–1049
Horsman A, Gallagher JC, Simpson M, Nordin BEC (1977) Prospective trial of oestrogen and calcium in postmenopausal women Br Med J ii:789–792
Hurxthal LM, Vose GP (1969) The relationship of dietary calcium intake to radiographic bone density in normal and osteoporotic persons. Calc Tiss Res 4:245–256
Irwin MI, Keinholz EW (1973) Conspectus of research on calcium requirements of man. J Nutr 103:1019–1095
Jeejebhoy KN (1986) Nutritional balance studies: indicators of human requirements or adaptative mechanisms. J Nutr 116:2061–2063

Johnson NE, Smith EL, Freudenheim JL (1985) Effects on blood pressure of calcium supplementation of women. Am J Clin Nutr 42:12–17
Kanders B, Dempster DW, Lindsay R (1988) Interaction of calcium nutrition and physical activity on bone mass in young women. J Bone Min Res 3:145–149
Kanis JA (1984) Treatment of osteoporotic fracture. Lancet i:27–33
Kanis JA, Passmore R (1989a) Calcium supplementation of the diet-I. Br Med J 298:137–140
Kanis JA, Passmore R (1989b) Calcium supplementation of the diet-II. Br Med J 298:205–208
Karanja N, McCarron DA (1986) Calcium and hypertension. Ann Rev Nutr 6:475–494
Kim KK, Kohrs MB, Grier MR (1984) Dietary calcium intakes of elderly Korean Americans. J Am Diet Assoc 84:164–169
Lau E, Donnan S, Barker DJP, Cooper C (1988) Physical activity and calcium intake in fracture of the proximal femur in Hong Kong. Br Med J 297:1441–1443
Leitch I, Aitken FC (1959) The estimation of calcium requirement: a re-examination. Nutr Abs Rev 29:393–411
MAFF (1970–1985) Household food consumption and expenditure. Annual reports of the National Food Survey Committee (15 volumes). Her Majesty's Stationery Office, London
Malm OJ (1958) Calcium requirement and adaptation in adult man. Scand J Clin Invest 10 [suppl 36]
Marie PJ, Pettifor JM, Ross FP, Glorieux FH (1982) Histological osteomalacia due to dietary calcium deficiency in children. N Engl J Med 397:585–588
Marshall DH, Nordin BEC, Speed R (1976) Calcium, phosphorus and magnesium requirement. Proc Nutr Soc 35:163–173
Massey LK, Hollingbery PW (1988) Acute effects of dietary content and assimilation on urinary mineral excretion in pre- and post-menopausal women. Nutr Res 8:845–851
Matkovic V, Kostial K, Simonovic I, Buzina R, Brodarec A, Nordin BEC (1979) Bone status and fracture rates in two regions of Yugoslavia. Am J Clin Nutr 32:540–549
M'Buyamba-Kabangu, Fagard R, Lijnen P, Bouillon R, Lissens W, Amery A (1987) Calcium, vitamin D-endocrine system and parathyroid hormone in black and white males. Calcif Tissue Int 41:70–74
Nielsen FH, Hunt CD, Mullen LM, Hunt JR (1987) Effect of dietary boron on mineral, Oestrogen and testosterone metabolism in postmenopausal women. FASEB J 1:394–397
Newmark HL, Wargovich MJ, Bruce WR (1984) Colon cancer and dietary fat, phosphate and calcium: a hypothesis. J Natl Cancer Inst 72:1323–1325
Newmark HL, Wargovich MJ (1986) Colon cancer and dietary fat, phosphorus and calcium in Hawaiian-Japanese men (letter). Am J Clin Nutr 44:998–1000
Newton-John HF, Morgan DB (1968) Osteoporosis: disease or senescence? Lancet i:232–233
Nilas L, Christiansen C, Rodbro P (1984) Calcium supplementation and postmenopausal bone loss. Br Med J 289:1103–1106
Nordin BEC, Marshall DH (1988) Dietary requirements for calcium. In: Nordin BEC (ed) Calcium in human biology. Springer-Verlag, Berlin, pp 447–464
Pansu D, Bellaton C, Bronner F (1981) Effect of Ca intake on saturable and nonsaturable components of duodenal calcium transport. Am J Physiol 240:G32–37
Pansu D, Bellaton C, Bronner F (1983) Developmental changes in the mechanisms of duodenal calcium transport in the rat. Am J Physiol 244:G20–26
Peacock M (1988) Renal excretion of calcium. In: Nordin BEC (ed) Calcium in human biology. Springer-Verlag, Berlin, pp 125–169
Pettifor JM, Ross P, Moodley G, Shuenyane E (1979) Calcium deficiency in rural black children in South Africa: a comparison between rural and urban communities. Am J Clin Nutr 32:2477–2483
Pettifor JM, Marie PJ, Sly MR, du Bruyn DB, Ross F, Isdale JM, de Klerk W, van der Walt WH (1984) The effect of differing dietary calcium and phosphorus contents on mineral metabolism and bone histomorphometry in young vitamin D-replete baboons. Calcif Tissue Int 36:668–676
Picard D, Ste-Marie LG, Coutu D, Carrier L, Chartrand R, Lepage R, Fugere P, D'Amour P (1988) Premenopausal bone mineral content relates to height, weight and calcium intake during early adulthood. Bone and Mineral 4:299–309
Polley KJ, Nordin BEC, Baghurst PA, Walker CJ, Chatterton BE (1987) Effect of calcium supplementation on forearm bone mineral content in postmenopausal women: a prospective, sequential controlled study. J Nutr 117:1929–1935
Recker RR, Saville PD, Heaney RP (1977) Effect of estrogens and calcium carbonate on bone loss in postmenopausal women. Ann Intern Med 87:649–655
Recker RR, Heaney RP (1985) The effect of milk supplements on calcium metabolism, bone metabolism and calcium balance. Am J Clin Nutr 41:254–263

Riggs BL, Wahner HW, Melton J, Richelson LS, Judd HL, O'Fallon WM (1987) Dietary calcium intake and rates of bone loss in women. J Clin Invest 80:979–982
Riis B, Thomsen K, Christiansen C (1987) Does calcium supplementation prevent postmenopausal bone loss? N Engl Med J 316:173–177
Royal College of Physicians (1989) Fractured neck of femur: prevention and management. The Royal College of Physicians, London
Sandler RB, Slemenda CW, LaPorte RE, Cauley JA, Schramm MM, Barresi ML, Kriska AM (1985) Postmenopausal bone density and milk consumption in childhood and adolescence. Am J Clin Nutr 42:270–274
Smith RW, Frame B (1965) Concurrent axial and appendicular osteoporosis. Its relation to calcium consumption. N Engl Med J 273:73–78
Spencer H, Kramer L, Osis D (1982) Factors contributing to calcium loss in aging. Am J Clin Nutr 36:776–787
Spencer H, Kramer L, Lesniak M, De Bartolo M, Norris C, Osis D (1984) Calcium requirements in humans. Clin Orthop Rel Res 184:270–280
Spencer H, Kramer L (1986) The calcium requirement and factors causing calcium loss. Fed Proc 45:2758–2762
Stevenson JC, Whitehead MI, Padwick M, Endacott JA, Sutton C, Banks LM, Freemantle C, Spinks TJ, Hesp R (1988) Dietary intake of calcium and postmenopausal bone loss. Br Med J 297:15–17
Truswell AS, Irwin T, Beaton GH, Suzue R, Haenel H, Hejda S, Hou X-C, Leveille G, Morava E, Pedersen J, Stephen JML (1983) Recommended dietary allowances around the world. A report by Committee 1/5 of the International Union of Nutritional Sciences 1982. Nutr Abs Rev 53:939–1119
Walker ARP (1954) Does a low intake of calcium retard growth or conduce to stuntedness? Am J Clin Nutr 2:265–271
Walker ARP (1972) The human requirement of calcium: should low intakes be supplemented? Am J Clin Nutr 25:518–530
Wardlaw GM, Pike AM (1986) The effect of lactation on peak adult shaft and ultra-distal forearm bone mass in women. Am J Clin Nutr 44:283–286
Yano K, Heilbrun LK, Wasnich RD, Hankin JH, Vogel JM (1985) The relationship between diet and bone mineral content of multiple skeletal sites in elderly Japanese-American men and women living in Hawaii. Am J Clin Nutr 42:877–888

Chapter 13

Magnesium: Clinical Forms of Primary Magnesium Deficiency

J. Durlach

Introduction

The ubiquitous character of magnesium and the diversity of its physiological functions explain the multiplicity of the clinical aspects of its deficit.

The best known and the earliest defined of the forms of primary magnesium deficiency (MDI) are the neuromuscular forms. In this chapter these will be described first and other aspects of MDI will then be considered more briefly: endocrine-humoral, cardiovascular, allergic and pseudo-allergic, gynaeco-obstetrical, osteo-articular, digestive, anaemic, infectious. The references at the end of this chapter provide more information about clinical forms of primary magnesium deficit.

The Neuromuscular Forms of MDI

Latent tetany (LT) due to chronic MDI represents the most commonly seen form of MDI in clinical practice. It will serve therefore as the descriptive model for neuromuscular hyperexcitability due to MDI.

Subjective Symptomatology

The symptoms leading to consultation include non-specific central, peripheral and autonomic manifestations absolutely identical to those previously described as latent tetany, hyperventilation syndrome, spasmophilia, Da Costa's syndrome, soldier's heart, effort syndrome, neurocirculatory asthenia and other names. The non-specific pattern of this symptomatology brings the patient to consult a wide range of specialists.

The neurotic, or rather, "central" symptoms consist of anxiety, hyperemotionality, fatigue, headaches (and sometimes migraine), insomnia, lightheadedness, dizziness, nervous fits, lipothymiae, sensation of a "lump in the throat", of "nuchalgia" and "blocked breathing".

The peripheral signs are acroparaesthesiae, cramps, muscle fasciculations and myalgiae.

The functional disorders include chest pain, sine materia dyspnoia, blocked respiration, precordialgia, palpitations, extrasystolae, dysrhythmias, Raynaud's syndrome, trends to orthostatic hypotension or conversely to borderline hypertension. In fact, the dysautonomic disturbances involve both the sympathetic and the parasympathetic systems.

When the chest pain mimics coronary heart disease, its relief with propranolol and worsening by nitrates may help to distinguish between a benign disorder and a trouble from coronary origin.

The evolution of LT may be studded with various acute paroxymal manifestations which can also sometimes be seen as initial signs of the illness. The major crises of acute tetany or of grand mal – even reduced to a simple loss of consciousness – remain relatively rare. It is more often a question of nervous crises: neurotic, from the "attack of nerves" to the "hysterical crises", or autonomic: lipothymia, reactive hypoglycaemia, pseudo-asthmatic crisis, vagovagal syncope or, on the contrary, paroxysmal tachycardia. Sometimes, centripetal tingling sensations and stiffness of the extremities confer on these nervous crisis a tetanoid character. But, essentially, they all have in common the fact that they occur in a context of fits of anxiety, even sometimes with the impression of imminent death (panic attack), which cause hyperventilation gaseous alkalosis and self perpetuation of the crises.

Physical Examination

On encountering this non-specific pattern the signs of neuromuscular excitability are of much greater importance.

A genuine Chvostek's sign must be systematically sought. With a small (children's) reflex hammer the examiner percusses the soft parts of the cheek at the centre of a line running from the ear lobe to the labial commissure, avoiding the lightning contraction of a "false Chvostek's sign" by tapping the bone of the zygomatic apophysis. It is important to consider the quality – and not the intensity – of this clinical criterion of neuromuscular hyperexcitability. It is only its presence or its absence which is significant, respectively quoted 1 or 0.

The examination of the precordial area will be carefully conducted in order to search either for a non-ejection systolic click, or for a mid- to end-systolic or pansystolic murmur, or both, particularly in orthostatism in complete expiration and in the left lateral decubitus position.

Tracings

Routine Tracings

Two tracings are always made: a neurophysiological examination (electromyogram (EMG)) and a cardiological examination (echocardiogram (ECC)).

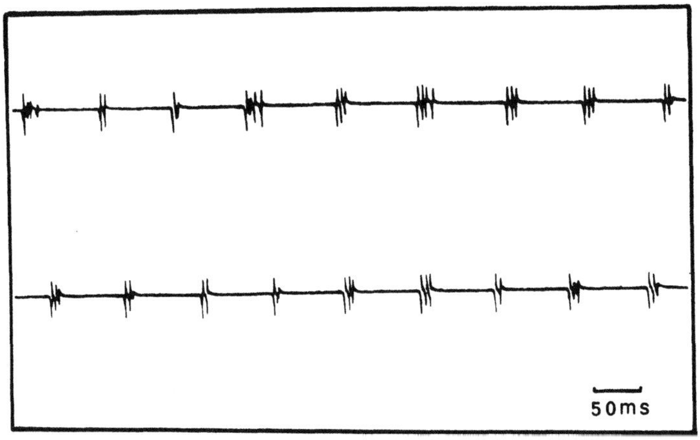

Fig. 13.1. Multiplets: an example of typical EMG repetitive activity observed during and after the removal of the tourniquet or at the end of hyperventilation (from Durlach 1988a)

In EMG testing for latent tetany, a Bronck's needle is inserted into the first dorsal interosseous muscle of the left hand. The three classical facilitation tests are used: tourniquet-induced ischaemia lasting 10 min, post-ischaemia lasting 10 min after removal of tourniquet and lastly hyperventilation lasting 5 min. If the EMG shows one (or several) train(s) of autorhythmic activities, "beating" for more than 2 min of one of the three tetanic activities (uniplets, multiplets or complex tonicoclonic tracing(s) a paritive reponse is defined. As determined for the clinical criterion of tetanic hyperexcitability, this neurophysiological criterion is only considered as a two-class variable. Either its presence or its absence is significant, respectively quoted 1 or 0 (Fig. 13.1).

The "excitability index" (EI) is defined as the sum of the two criteria of tetany. It allows different classes among tetanies to be distinguished: one with simultaneous clinical and neurophysiological criteria: EI=2, the others with only one criterion of their tetanic state: EI=1 with two sub-groups, either clinical (through positivity of the Chvostek's sign alone), or electromyographic (through positivity of EMG alone).

The ECC is the best tool for detecting mitral valve prolapse (MVP). With time-motion (TM) mode, three tracings are classical: a "cuplike" tracing of mesotelesystolic MVP (of more than 2 mm) (Fig. 13.2), a "hammocking" tracing of holosystolic MVP (of more than 3 mm) (Fig. 13.3), an isolated systolic anterior motion (SAM) (Fig. 13.4) observed without obstruction nor any septal thickening sign and in the absence of false systolic anterior motion.

Two-dimensional echocardiography appears to be more accurate than TM echocardiography. It eliminates a number of artefacts and, particularly, in the

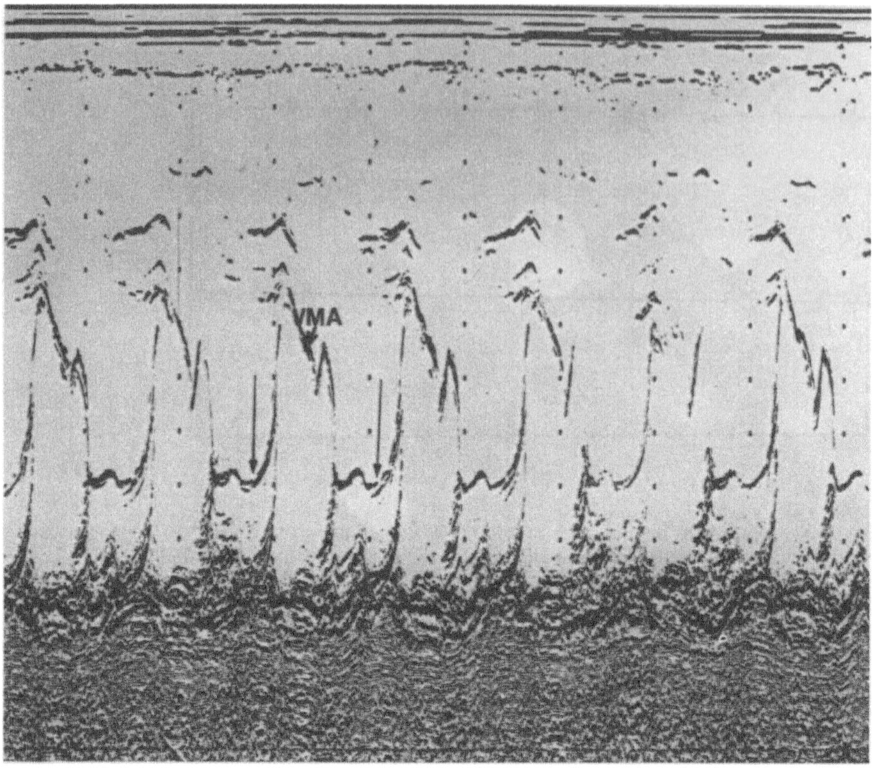

Fig. 13.2. A time motion echocardigram shows a cup-like aspect in the case of mesotelesystolic MVP (from Durlach 1988a)

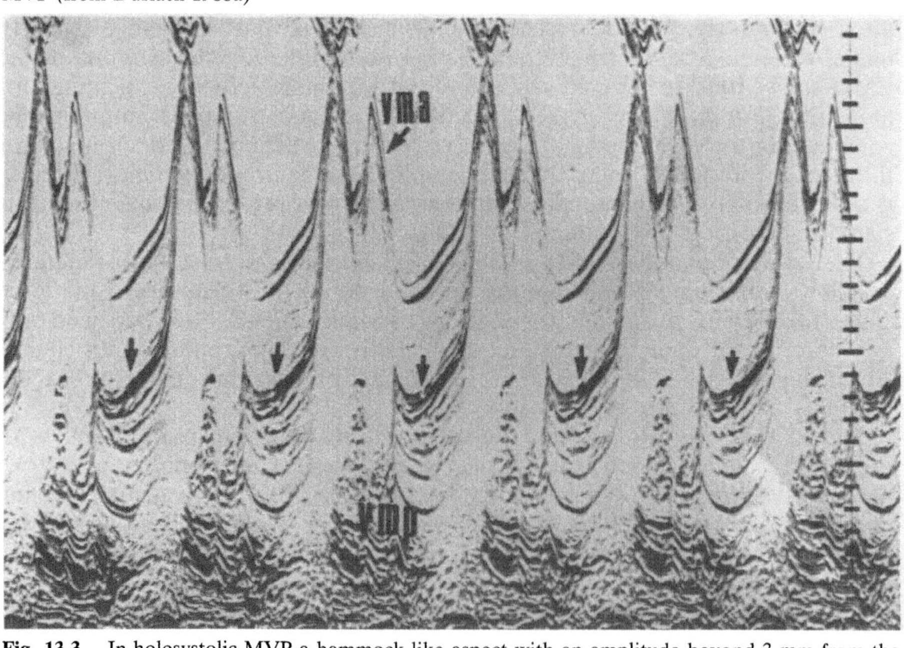

Fig. 13.3. In holosystolic MVP a hammock-like aspect with an amplitude beyond 3 mm from the CD line, with TM mode (from Durlach 1988a)

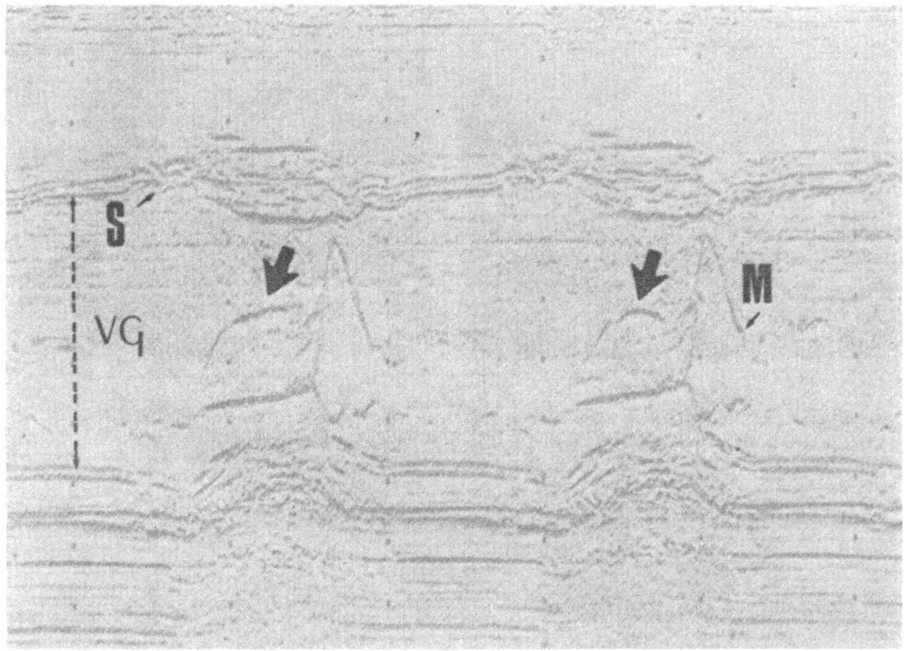

Fig. 13.4. An isolated systolic anterior motion (SAM) without any sign of obstruction, or any septal thickening, corresponds to a genuine MVP, in TM mode (from Durlach 1988a)

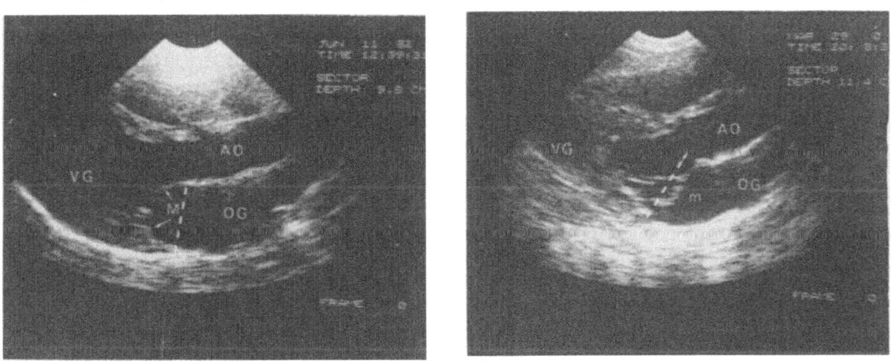

Fig. 13.5. The parasternal cross-section allows adequate observation of mitral kinetics, in two-directional mode (from Durlach 1988a). AO, aorta; OG, left auricle; M, mitral valve; VG, left ventricle (from Durlach 1988a).

section of parasternal longitudinal cut (Fig. 13.5) and the apical cut of the four heart chamber (Fig. 13.6). The criterion for mitral prolapse is the billowing of one or both leaflets below the level of the mitral ring. It is very important to assess the leaflet thickness as well as its whole morphology and to appreciate the ventricular kinetic by calculating:

$$\triangle D = \frac{\text{end diastolic diameter} - \text{end systolic diameter}}{\text{end diastolic diameter}}$$

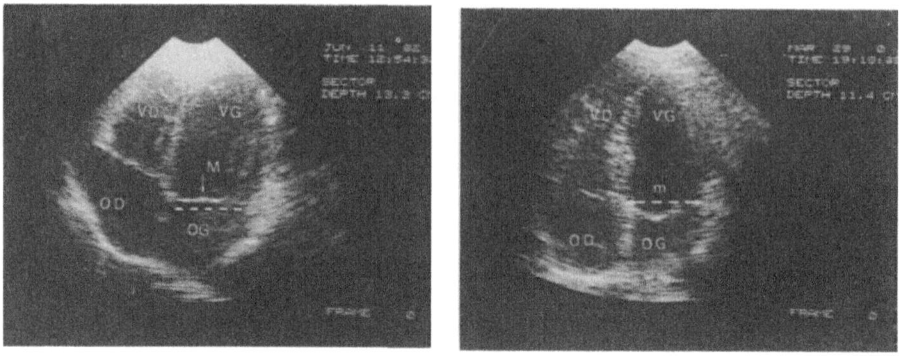

Fig. 13.6. The apical cross-section allows adequate observation of the mitral and tricuspid valve kinetics in two-dimensional mode (from Durlach 1988a). M, mitral valve; OD, right auricle; OG, left auricle, VD, right ventricle; VG, left ventricle.

Pulsed doppler echocardiography allows the detection of associated mitral regurgitation.

In Particular Clinical Forms

In particular forms, more sophisticated neurophysiological or cardiological explorations may be useful.

Additional neurophysiological examinations can be used in the evaluation of the psychic and neuromuscular condition, i.e., standard electroencephalogram, electropolygraphic study of light sleep, electronystagmography, optokinetic test, head scan and psychometric investigations, for example: Minnesota Multiphasic Personality Inventory (MMPI) or evaluation of A/B behaviour pattern by Jenkin's questionnaire.

Additional cardiological examinations may allow a better appreciation of the cardiac condition, i.e., standard and/or effort electrocardiogram, ambulatory electrocardiogram (Holter), phonocardiogram, or, very exceptionally, angiocardiogram.

Ionic Evaluation

Routine Ionic Assessment

Four ionic investigations should always be made: plasma Mg (pMg), erythrocyte Mg (eMg), calcaemia and daily calciuria, which can be completed by the finding of urinary infection. These must first demonstrate normocalcaemia and the absence of hypercalciuria susceptible to induce a secondary magnesium deficit. Next, the evaluation of pMg and eMg with reliable methods, such as atomic absorption spectrophotometry, allows the diagnosis of primary magnesium deficit through hypomagnesaemia in one-third of the cases of latent tetany due to MDI, with or without MVP (Table 13.1). Normal levels do not rule out the diagnosis of MDI. The histograms of LT patients (with or without MVP) and of controls overlap. If the tetanic group reveals gaussian-type magnesaemia curves with

Table 13.1. Normal and hypomagnesaemic levels of plasma and erythrocyte magnesium, from Durlach (1988a). Values are given as means ± 2SD. The threshold value for plasma hypomagnesaemia is 0.8 mmol/l (19 mg/l) and for erythrocyte hypomagnesaemia it is 1.8 mmol/l (44 mg/l)

	Controls				Hypomagnesaemic patients	
	Plasma		Erythrocyte		Plasma	Erythrocyte
	Mean	SE	Mean	SE	Mean – 2SE	Mean – 2SE
mmol/l	0.88	0.05	2.30	0.24	0.78	1.82
mg/l	21.40	1.14	55.94	5.96	19.12	44.02

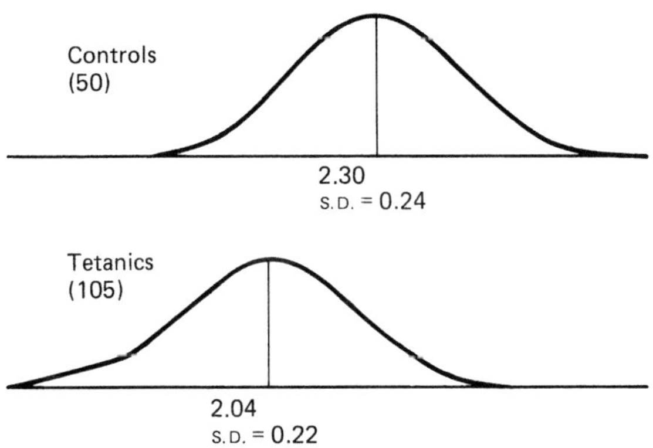

Fig. 13.7. Erythrocyte magnesium levels in a population of patients with latent tetany due to primary magnesium deficiency and in a control group. The measurements were obtained with atomic absorption spectrophotometry. There is an overlapping between the two curves, the curve for the tetanic patients being shifted significantly to the left. Its mean is significantly lower (P>0.001). Depending on the position from left to right on the curve, the patient with tetany due to magnesium deficiency can have low (⅓ of cases), normal (about ⅔ of cases) or high (rare) magnesium levels. (From Durlach 1988a.)

significant lower means (P<0.001) both for pMg and eMg, their constitutive elements can be individually hypomagnesaemic (one-third of the cases), normomagnesaemic (almost two-thirds of the cases) and even, although seldom, hypermagnesaemic (Fig. 13.7). Nevertheless one must emphasize the remarkable constancy of magnesaemia which lends importance even to small variations of magnesaemia.

In numerous cases (particularly those with normal parameters) the diagnosis requires a magnesium oral loading test. The dose of magnesium to be administered is 5mg/kg/day of a well absorbed salt for at least one month. At this physiological dose level, oral magnesium is totally devoid of the pharmacodynamic effects of parenteral magnesium. Correction of symptoms by this oral magnesium load constitutes the best proof that they were due to magnesium deficiency, as checked after one month of treatment.

However, a negative response does not permit rejection of a diagnosis of magnesium deficit. A mere increase of the magnesium intake is sometimes inadequate to ensure uptake or maintenance of cellular magnesium, in forms of deficit corresponding to magnesium depletion due to a dysregulation of the factors which control its metabolism. Therefore, if the response remains negative after one month's physiological magnesium oral loading test, it should be followed by use of the best specific control of the magnesium metabolism dysregulating factors. Otherwise the investigation may be continued with the addition of a non-specific magnesium fixing agent to the magnesium salt: for example pharmacological doses of vitamin B6, physiological doses of vitamin D metabolites or of Mg-sparing diuretics able to reduce a possible hypermagnesuria (Table 13.2).

Table 13.2. Comparisons between oral and parenteral magnesium load tests

Method of administration	Oral	Parenteral
Dose	Physiological	Pharmacological
Administration	Ambulatory	In hospital
Criteria of evaluation		
Clinical	+ + +	0
Biological	+ + +	+ + +
Duration	Several weeks	4 hours
Indications		
Deficiency	+ + +	+ + +
Depletion	+	0

Thus pMg, eMg, calcaemia and calciuria are the first minimum ionic measurements required. They are often completed by checking daily magnesuria in order to control the effectiveness of the magnesium oral loading test after the first month of treatment.

In Particular Clinical Forms

More sophisticated studies may be carried out to determine magnesium or other metabolites. Some studies are static: dietary inventory, magnesium balance, ionized and diffusible fractions of plasma magnesium, levels of magnesium in cerebrospinal fluid, lymphocytes, bone, muscle, even platelets and exceptionally myocardium. Other investigations are dynamic. The most useful is the parenteral magnesium load test: for example, Thoren's technique, the efficiency of which is judged by the modifications of the magnesuria four hours after the infusion of 0.25 mM (about 6 mg)/kg/day. Thus a retention which exceeds 20% of the magnesium supplied, corresponding to a magnesuria lower than 80% of the parenteral load, supports the diagnosis of magnesium deficiency (positive Thoren test). This parenteral load test serves as a rapid diagnosis of deficiency, but it can only be carried out in hospital. As the clinical effects of parenteral pharmacological doses might be purely pharmacodynamic, the results can only be assessed on the basis of objective criteria. Finally a negative Thoren's test may be observed in cases of depletion ("false negative response" for this form of magnesium deficit), as well as in cases of a balanced metabolism ("true negative response") (Table 13.2).

Evaluation of modifications of the magnesaemia during a standardized effort (a treadmill test and/or a mental task), after glucose or calcium or phosphorus loads and, exceptionally, radioisotopic explorations may complete investigations into magnesium balance. Other metabolites may be simultaneously explored in plasma, erythrocyte, blood, urine, such as potassium, zinc, iron, phosphorus, phosphatases, ATP, creatine, creatine kinase, glucose, aminoacids, essential fatty acids.

Complete Record

In rare cases, it may be interesting to evaluate both the neuro-endocrine factors of magnesium homeostasis and the main disrupting elements. These include on the one hand: measurements of adrenaline, noradrenaline, PTH, CT, insulin, taurine, cyclic AMP and GMP and/or, on the other hand, measurements of angiotensin, renin, aldosterone, vitamin D metabolites, free T3 and T4. Sometimes, it may be useful to investigate immunological data such as immunoglobulin levels, histamine, acetylcholine, noradrenalin, isoprenaline receptivities, and haemorrheological data such as platelet and erythrocyte explorations, familial data (history, HLA typing etc.).

Evolution and Prognosis

Usually, the prognosis of the neuromuscular form of MDI is favourable. The evolution progresses by phases of decompensation against a background of latency.

It appears of major importance to distinguish between the larger group of patients (95% of the cases) with a benign prognosis and even with the possibility of complete recovery, and the smaller subgroup of patients (5% of the cases) running the risk of complications. In this latter population, MD appears as a nervous and cardiovascular risk factor more exposed to arrhythmias, sudden death, endocarditis, cerebral or visual ischaemic events.

Among the main factors of a favourable prognosis are latency or paucity of clinical and paraclinical symptomatology: i.e., low excitability index (IE=1), absence of auscultatory signs, non-redundant and thin leaflets, absence of mitral regurgitation, normal weight, no oestrogen intake and lastly the deficient type of the magnesium deficit with its simple treatment.

Conversely, it is important to determine some pejorative prognosis factors, i.e. a rich symptomatology with ventricular arrhythmias particularly, a high excitability index (IE=2), not so much a click as mitral regurgitation stigmata, redundant thick mitral leaflets especially in older men, prolongation of QTc interval, underweight, thrombogenic disturbances mainly through alterations of the platelet function, immunological disorders, constitutional factors such as the carrying of the HLA Bw35 antigen, exceptionally a familial history of sudden death, lastly the depletion type of the magnesium deficit with its often difficult, and sometimes chronic, treatment.

Table 13.3. The manifestations of primary magnesium deficit (Durlach 1988a). Those conditions listed in capital letters are the forms known to exist and those in lower case letters are probable but not yet proven to occur. The conditions in bold type are the most frequently occurring, and those in italics merit special attention

NEUROMUSCULAR
LATENT TETANY (spasmophilia, hyperventilation syndrome)
  With or without **MITRAL VALVE PROLAPSE** (Da Costa's syndrome, irritable heart, neurocirculatory asthenia)
  Large female prevalence
NON-TETANIC NEUROMUSCULAR FORMS
  Uncomplicated by mitral valve prolapse
  Smaller female prevalence
  Rare complicated forms (ISOLATED MITRAL INSUFFICIENCY, RHYTHM DISORDERS (lengthened QT), ischaemic attacks

CARDIOVASCULAR
Megnesium deficit "cardiovascular risk factor"
  Arterial hypertensions, labile or permanent, hypercalcaemic, hyperreninaemic, hypophosphoraemic
PHLEBOTHROMBOGENIC FORM

ENDOCRINE HUMORAL
HYPOCALCAEMIA (*vitamin resistant*)
  Hypoparathyroidism
  Pseudohypoparathyroidism
  Osteomalacia or rickets
  Hypocalcaemic cardiomegaly
HYPOKALAEMIA
  Without juxtaglomerular hypertrophy
  With juxtaglomerular hypertrophy (Bartter's syndrome due to primary magnesium deficit)
REACTIVE HYPOGLYCAEMIA
  Hypercalcaemic forms
  Hypermagnesemic forms
  Diabetic forms

ALLERGIC
Type I allergies, anaphylactic shock,
  often associated with latent tetany, especially in its pseudo-allergic form

OSTEO-ARTICULAR
Bones: Vitamin-resistant *rickets and osteomalacia*
  Growth retardation
Joints: *CHONDROCALCINOSIS*

OBSTETRIC
  Spontaneous abortion
  Premature birth
  Dysgravidia
  Fetal hypotrophy and fragility

DIGESTIVE
Transit disorders: *constipation, ileus, diarrhoea, steatorrhoea*

ANAEMIC
Especially in association with protein malnutrition, iron, B6 and B12 deficiencies and lead poisoning

INFECTIOUS
Reduced resistance to infection

Other Clinical Forms of MDI (Table 13.3)

As is the case with any ionic deficit, magnesium deficiency can, depending on individual susceptibility, relate to these targets collectively or rather dominate or not be expressed in any one of them. Since the principal form of expression of MDI remains its psychic, neuromuscular and autonomic symptomatology, one must always then seek out, in association with other, rarer manifestations of MDI the evidence of a latent tetany due to the deficiency.

Endocrine-humoral Forms

It is proper to distinguish, among the endocrine-humoral forms of MDI, between those of which the identification is certain and those for which the description still lacks certain definitive features.

Among the endocrine-humoral forms of certain identity, it is possible to observe mainly hypocalcaemic, hypokalaemic or hypoglycaemic forms.

MDI presents three clinical aspects of hypocalcaemia: those of hypoparathyroidism, or pseudohypoparathyroidism or of vitamin D deficiency. The clinical aspects of hypokalaemia are either simple hypokalaemia without juxtaglomerular hyperplasia, or Bartter's syndrome with juxtaglomerular hyperplasia and high blood renin.

The hypoglycaemic forms may or may not be accompanied by convulsions. The hyperphosphaturic forms may be associated with lithiasis. Hyponatraemic forms are found infrequently.

All these cases are resistant to their specific treatments, but are curable by treatment with magnesium. In the present state of our knowledge the individuality of the other humoral forms remains uncertain.

Nevertheless, their existence is probable and a possible magnesium aetiology should be considered when confronted with type I diabetes or hypophosphatasaemia or hypercalcaemia. It is particularly interesting to point out the existence of a clinical form of hypertension with hypercalcaemia, hypophosphoraemia and high blood level of renin. It will be magnesium curable.

Lastly, high levels of magnesium may be observed in plasma during MD.

Cardiovascular Forms

Magnesium deficit and particularly its deficiency in water constitutes a cardiovascular risk factor. It has been supported by numerous well-documented epidemiological, experimental and clinical data. It is one of the factors predisposing to atherosclerosis, ischaemia, arrhythmia and sudden death. But, sometimes, it is only a cofactor in the genesis of hypertension. Its action on blood pressure is more often null or even hypotensive. The frequency of its aetiopathogenic role in "idiopathic" phlebothrombosis is still uncertain.

Allergic and Pseudo-allergic Forms

The MDI may be revealed by Type 1 allergic episodes. Allergic forms of MD are either acute; such as anaphylactic shock, or chronic such as reactive allergies which induce asthma, rhinitis, conjunctivitis, urticaria, Quincke's oedema and migraines.

The pseudo-allergic form is usually associated with latent tetany and is due to histamine and acetylcholine hyperreceptivity. It is observed either alone, or with the allergic form.

A superacute form of MD may intervene in the genesis of sudden death, both of neonates and adults, and particularly during anaesthesia.

Gynaeco-obstetrical Forms

If in gynaecology, MDI may have a part in the genesis of "essential" dysmenorrhoea, it plays an important role during pregnancy both with the mother and the fetus. It predisposes to spontaneous abortions, premature labour and birth. Its aetiopathogenic role is uncertain in eclampsia.

Magnesium is necessary for adequate development and trophicity of the fetus.

Osteo-articulary, Digestive, Anaemic Forms and Infection

MD may play a role in some chondrocalcinosis, growth retardation, vitamin D-resistant rickets and osteomalacia, in digestive transit disorders: sometimes constipation and even ileus, sometimes, on the contrary, diarrhoea, steatorrhoea, anaemias, usually in association with other promoting factors such as protein malnutrition, Fe or B6 or B12 deficiencies, and lead poisoning. Lastly MD may reduce the resistance to infection.

Conclusion

Latent tetany with or without mitral valve prolapse and allergies with or without pseudo-allergy represent by far the most frequent expressions of MDI. Particular attention should be given to various humoral aspects with hypocalcaemia, hypokalaemia and hypoglycaemia and to the frequency of MDI in difficult pregnancies, some types of "essential" dysmenorrhoea, chondrocalcinosis, growth retardation, and "idiopathic" phlebothrombosis. Such a list is only provisional. Certain forms of anaemia, certain digestive problems and low resistance to infection will probably find their proper place among the clinical forms of the primary magnesium deficit since in our present state of knowledge it seems difficult to take sufficiently into account the ubiquitous nature of Mg and the multiplicity of its physiological functions.

Since the principal form of expression of MDI remains LT due to MD one must always then seek out its manifestations: the diagnosis must depend on Mg oral loading tests conducted with physiological doses.

References

Durlach J (1985) Neurological disturbances due to magnesium imbalance. In: Gabay S, Harris J, Ho BT (eds) Metal ions in neurology and psychiatry. Alan R Liss, New York, pp 121–128

Durlach J (1988a) Magnesium in clinical practice. John Libbey, London

Durlach J (1988b) Le métabolisme du magnésium et son rôle en pathologie. In: Encycl Med Chir Glandes-Nutrition 10357 A^{10} 5, p 8

Durlach J (1988c) Les rapports entre le magnésium et le prolapsus de la valve mitrale. Rev Méd Fonctionnelle 20:021–172

Durlach J, Durlach V (1986) Idiopathic mitral valve prolapse and magnesium. State of the art. Magnesium-Bull 8:156–169

Durlach J, Durlach V, Poenaru S, Rouhani S (1985) Physiologic tracings and ionic evaluation of latent tetany due to magnesium deficit. J Am Coll Nutr 4:333

Durlach J, Poenaru S, Rouhani S, Bara M, Guiet-Bara A (1987) The control of central neural hyperexcitability in magnesium deficiency. In: Essman WB (ed) Nutrients and brain function. Karger, Basel, pp 48–71

Durlach J, Bara M, Guiet-Bara A (1989) Magnesium levels in drinking water: its importance in cardiovascular risk. In: Itokawa Y, Durlach J (eds) Magnesium in health and disease. John Libbey, London, pp 173–182

Chapter 14

The Functional Significance of Iron Deficiency

L. Hallberg

Introduction

Iron is an essential element present in all cells. Iron-containing compounds have several vital functions such as the transportation of oxygen to the cells and electrons within the cells. In iron deficiency, there is usually a reduction in most of these compounds. Much more is known about the marvellous biochemical achievements performed by these compounds and about the ingenious systems regulating their functions than about the negative effects encountered in states of iron deficiency.

Until recently, few firm facts were known about the effects and symptoms in man that could be related to iron deficiency. In the last 15 years, however, much research has been devoted to this field and several excellent reviews have been published (Dallman 1986, 1987; Lozoff and Brittenham 1986; Pollitt et al. 1986; Scrimshaw 1984).

The purpose of this paper is to summarize some of the main findings and to outline some of the areas where more knowledge is needed.

The Main Iron Compounds in the Body and Their Function

Functional Iron Compounds

Haemoglobin comprises the main part of body iron. An adult male has about 2400 mg iron in the form of haemoglobin, about 65% of the total amount of iron. The main function is to transport oxygen to the tissues from the lungs.

Myoglobin stores oxygen in the muscle for use during muscle contraction. It is a monomeric unit of haemoglobin, with one globin chain and one haem group with one iron atom. (Haemoglobin is composed of 4 units.) About 5 mg myoglobin is present per gram of tissue and constitutes about 10% of the total body iron.

Cytochromes are a series of intracellular enzymes which also contain a haem group engaged in the transport of electrons, mainly in the mitochondria. These enzymes are essential for oxidative phosphorylation and the formation of ATP and thus for the generation of cellular energy.

There are several *other haem-containing enzymes* such as cytochrome P-450, present especially in the liver microsomes and adrenal mitochondria, and engaged in the detoxication of foreign substances by hydroxylation and in the synthesis of various compounds such as steroid hormones and bile salts. Peroxidases, such as the important myeloperoxidase in the granulocytes and catalases, are other examples of haem enzymes.

Non-haem iron enzymes form another important group of iron compounds. Metalloflavoproteins and iron-sulphur proteins are engaged in oxidative metabolism and comprise enzymes such as NADHdehydrogenase, xanthine oxidase and succinic dehydrogenase.

Several *other enzymes* require iron as an activator or cofactor.

Storage Iron Compounds

The body has specially designed proteins for storage of iron, *ferritin* and *haemosiderin*. Inside the cavity of the ball-shaped ferritin protein, as many as 4500 ferric ions can be stored. Ferritin molecules may be converted to haemosiderin. Under normal conditions, about half of the storage iron is present as haemosiderin. Iron can be mobilized from both compounds, but probably faster from ferritin. Iron storage proteins are mainly found in the reticuloendothelial cells in hepatocytes and erythroid precursors. The main part of iron is thus stored in the liver. The total normal amount of stored iron varies from zero to about 1500 mg. In women, the stores seldom amount to more than 500 mg; in men, iron stores successively accumulate to about 1000–1500 mg, depending on age and diet. The main physiological role of iron stores is probably to serve as a "bank account" to be used during pregnancy when normal iron balance cannot be maintained by diet alone.

Transferrin

Transferrin binds two atoms of ferric iron and is a protein specially designed for the transport of iron between different compartments in the body. Its main destinations are red cell precursors, which account for about 80% of the plasma iron turnover. All growing cells in the body, however, receive their iron from transferrin. A special transferrin receptor on the surface of the cells is needed for the active transfer of the iron. Only a very small amount of iron (about 3 mg) in the body is present in this transport form. In a state of iron deficiency and in pregnancy, the concentration of transferrin increases by an increased synthesis in the liver, probably in order to facilitate absorption i.e. the transport from intestines to other tissues.

Iron Metabolism

Iron metabolism can be described as two loops – one internal loop with a continuous re-utilization of iron released from cells catabolized within the body (mainly from haemoglobin in red cells), and one external loop represented by the absorption of iron from the diet and the losses of iron from the body (mainly from cells lost from the body's external and internal surfaces) (Hallberg 1982).

Development of Iron Deficiency

A negative iron balance develops when the absorption of iron cannot cover the losses of iron from the body or the demands for growth and pregnancy. If this development is slow, there is a successive depletion of iron stores, if present. The body responds by increasing the absorption of iron in an attempt to maintain iron balance. This may succeed – if not, the negative iron balance continues. The transferrin concentration in plasma increases and the plasma iron level decreases, leading to a reduction of iron on the two binding sites for iron on the transferrin molecule (reduced transferrin saturation). A consequence will be that less iron reaches the red cell precursors in the bone marrow and that the synthesis of haemoglobin is impaired. The haemoglobin concentration (and later the number of red cells) will start to decrease. It is easy to measure the haemoglobin concentration accurately. In normal subjects, the levels are very constant but vary markedly between different individuals. A low haemoglobin level in an individual, even if it is quite marked, cannot be diagnosed as probably pathological, as anaemia, until the level has fallen below the lower 95% level for haemoglobin in a male or female *population*, i.e., below 120 g/l in a women who may have 150 g/l as her normal level.

At about the same time as the iron supply to red cell precursors is impaired, the supply of iron to other tissues will be threatened. The development of a deficiency in other tissues depends on many factors. It is probable that the supply of iron to a cell only occurs during an early phase of the cell's life cycle as in the case for red blood cells. The supply of iron to a tissue will thus be determined by time, effective plasma flow, the transferrin-iron saturation and the number of active transferrin receptors on the cell surface, which in turn may be related to the turnover rate of the cells.

This would explain the observations that the content of cytochromes in intestinal cells with a life span of only a few days is more rapidly reduced than the haemoglobin in red cells with a life span of 120 days and that the repair is much faster in intestinal cells when iron is given.

Actually, this dynamic concept of the iron situation in the body implies that tissue deficits may occur without a person being anaemic (as usually defined); that some of the negative effects of iron deficiency will only be found in subjects with iron deficiency of long duration due to a low cell turnover rate; and that possibly some negative effect may only be found in subjects who were iron-deficient in childhood and adolescence i.e., during growth and organ development. This latter effect may be related to some of the symptoms of impaired brain function e.g., learning. What is said about the development of symptoms in a state of iron deficiency must also be valid for the effect of treatment. It is reasonable to assume that the reversibility of symptoms is related to the rate of renewal of deficient cells and that some symptoms are, therefore, resistant to treatment with iron. The main point to consider is that negative findings of the effects of iron deficiency or effects of treatment in acute experiments cannot be used as counter-evidence for a causal relationship between iron deficiency and certain symptoms.

There are several *other confounding factors* in the analysis of the relationship between iron deficiency and body functions. The following are some examples.

A reduction of the content of certain iron-containing enzymes in a tissue may not necessarily be accompanied by any reduction in function if the enzyme is not rate-limiting for a certain metabolic process. Many enzymes are actually present in excess in many tissues. Another type of problem is analysing causality between iron deficiency and symptom can be that the iron deficiency is only one facet of general malnutrition and that some other concomitant deficiency is responsible for certain effects. These kinds of confounding factors in the relationship between iron deficiency and brain function have recently been reviewed (Hallberg 1989).

Compensatory Mechanisms

The compensatory adaptations that occur in anaemia are a good example of nature's ability to adapt to a deficiency and to a reduced amount of oxygen-carrying haemoglobin. The first measure is that more oxygen is extracted from the haemoglobin in the tissues. As more lactic acid is formed, more oxygen is released from haemoglobin due to a change of the form of the oxygen dissociation curve (Bohr effect). An increased content of reduced haemoglobin in the red cells induces an increased level of 2-, 3-diglycerophosphate, which in turn influences the haemoglobin–oxygen affinity so that more oxygen is released. Increase of cardiac output and redistribution of blood flow to vital organs (e.g., brain, heart) further reduces the probability of the development of symptoms. In fact, even patients with severe anaemia may not complain of any symptoms if the anaemia has developed slowly and they have a sedentary lifestyle and are otherwise healthy.

Effects of Iron Deficiency

In spite of the difficulty in measuring the negative effects of iron deficiency, there are several studies which demonstrate or support a causal relationship between iron deficiency and impairments of function.

Work Capacity

In 1972, Edgerton et al. reported that work performance in rats was impaired by anaemia and that iron treatment normalized the situation within three to four days. In 1976, Finch et al., in a series of ingenious experiments, showed that the impairment of work performance in iron deficient rats was not due to the anaemia per se, but to a tissue-iron deficiency which could be restored by iron before anaemia had returned to normal. It was then shown that exchange transfusion to a normal Hb level could almost normalize the VO_2max and the ability to perform short intense exercise, whereas endurance exercise was almost unaffected by exchange transfusion. We know now that fast-twitch fibres, which are responsible for rapid movements, are poor in oxidative enzymes and are dependent on glycolysis and oxygen in myoglobin as energy sources, whereas slow-twitch muscles have 3–5 times the oxidative capacity and are rich in oxidative enzymes. Several further studies in animals strongly suggest that iron deficiency is associated with an impairment of physical endurance activity (Davies et al. 1982, 1984).

Studies in Guatemala by Viteri et al. (Viteri and Torun 1974) were the first to demonstrate a relationship between iron deficiency anaemia in agricultural workers and a test of physical work capacity (Harvard step test). Of special importance in this study was the observation that there was a relationship between haemoglobin level and submaximal work capacity, not only at low haemoglobin levels, but well up within the normal haemoglobin range. Several studies were then undertaken to investigate further the relationship between iron deficiency and work performance.

One study was carried out in Sri Lanka on a tea plantation (Gardner et al. 1977; Edgerton et al. 1981) and two other studies in Indonesia (Basta 1974, Husaini et al. 1983). For a review of these studies, see Schrimshaw (1984). They illustrate the difficulty in measuring actual work output and in designing and running these kinds of studies with adequate control groups, run-in periods etc. Some of the results are difficult to interpret and do not provide conclusive evidence for a causal relationship. The results strongly suggest, however, that there is a causal relationship.

Even if some of the results are difficult to interpret and do not provide clear, conclusive evidence of a causal relationship, the findings nevertheless strongly suggest that iron deficiency impairs work performance and productivity. This interpretation is also supported by the usually very concordant findings in animal studies where the experimental conditions are so much easier to control (Dallman 1986).

Effects on Brain Function (Youdim 1988)

Certain areas and structures of the brain, especially the basal ganglia (globus pallidus, red nucleus and substantia niger), have a content of non-haem iron of the same magnitude as found in major iron storage sites (liver, spleen and bone marrow). At birth, the brain iron concentration is only about 10% of that in adults. There is a gradual increase in brain iron and, at 10 years of age, the level is about half that found in adults. The functional importance of brain iron is supported by the finding of transferrin receptors for the transport of iron into the brain (Jefferies et al. 1984). The distribution of iron in the brain suggests an association with the metabolism of neurotransmitters and/or related compounds.

Dallman et al. (1975) showed that dietary iron deficiency in young rats led to a reduced content of hon-haem iron in the brain. Iron treatment corrected the anaemia and normalized the content of iron in liver, but brain iron could not be replaced. This has been confirmed in several studies. A deficit in brain iron persists in adult rats. These studies thus show that an insufficient supply of iron to a young animal leads to a lower content of brain iron, which is difficult to restore and thus remains in the adult animal (Dallman 1977).

Much work has been done in recent years to examine the significance of iron deficiency for brain function. An early observation was that the catabolism of parenterally administered monoamines was decreased in the non-deficient rat (Symes et al. 1971). Later, it was found that the synthesis of monoaminoxidase (MAO) was impaired (Youdim et al. 1975). It was also early observed that iron-deficient rats showed less spontaneous activity and a reversal of the normal diurnal pattern of activity (Glover and Jacob 1977). These findings have been confirmed in a series of studies (Youdim and Green 1977). Another early finding was that children with iron-deficiency anaemia had increased amounts of noradrenalin in urine, which was normalized within a week of iron treatment (Voornees et al. 1975). Much of the research on iron and brain function now became focused on its relationship to biogenic amines and the function of neurotransmitters.

Important findings in iron deficiency were a marked impairment of the binding of dopamine to the D_2 receptor of the caudate nucleus and of serotonin (5HT) to brain-synaptic vesicles (Ashkenazi et al. 1982). The changes were not related to anaemia but to iron deficiency, and were only partly and slowly reversible by iron treatment (Kaladhar and Rao 1982). The circadian rhythm pattern in motor activity, pain threshold, and thermoregulation are also dopamine mediated and reversed in iron-deficient rats (Youdim and Yehuda 1985).

There are experimental data to suggest that endogenous opiopeptides are involved in memory and learning and that there is a direct link to the dopaminergic neurotransmission, and thus between iron deficiency and learning (Yehuda and Youdim 1984). It was recently found that iron-deficient rats show cognitive deficits in learning and in memory tasks that involve attention (Yehuda et al. 1988).

There is thus a growing mass of knowledge in animals which strongly suggests links between changes in brain iron content, with changes in neurotransmitter systems and effects on behaviour and learning (Weinberg et al. 1980, 1981).

The relationship between iron deficiency and behaviour is much more difficult to study in man, since several methodological pitfalls must be considered to avoid serious bias factors. Poor socio-economic factors and/or genetic factors may influence both the development of a child, the nutritional situation, prevalence of infections etc., and the development of iron deficiency. Researchers in this field, however, have become increasingly aware of these problems. Most studies compare iron-deficient and control subjects before and after iron treatment, which should be designed as a double blind trial. An important and unavoidable problem is that some symptoms (measurements) may not be reversible by iron treatment or may require very long iron supplementation to achieve an improvement. A main cause is the slow turnover of many cells in the central nervous system. There are thus systematic errors that may both underestimate and overestimate a true causal relationship between iron deficiency and brain function. A further problem is that the tests used are fairly crude and have a low specificity and sensitivity. This implies that only relatively large changes may reach statistical significance and that hypothetical true and important changes in low-grade iron deficiency may not be detected.

Several studies in infants have shown significant behavioural differences between those with iron deficiency anaemia and non-iron-deficient infants (Lozoff et al. 1982a,b; Walter et al. 1983). Even if great care has been taken to make the groups fully comparable with respect to several factors, the effects of differences due to stimulation in the homes and the mothers' IQ cannot be excluded (Lozoff 1988). Studies using a randomized, placebo-controlled, double-blind experimental design have failed to show an effect of iron on mental performance for up to three months. It should be pointed out, however, that compliance has not been good in all studies.

These behavioural studies in infants were recently critically discussed in reviews (Lozoff and Brittenham 1986; Lozoff 1988), which indicated that several well-designed studies have consistently shown lower mental scores in infants with iron deficiency anaemia, which may indicate a causal relationship. The possibility that iron deficiency may cause irreversible brain damage is supported by several studies in rats, where low brain iron and behavioural effects persisted into adulthood despite correction of the anaemia. A similar conclusion may be drawn from a study on infants who were anaemic at nine months and who showed a lower IQ and developmental test up to four years later, after controlling for birth-weight and socio-economic factors (Palti et al. 1983).

The results of many studies do not prove, but do suggest, that there may be lasting ill-effects of iron deficiency in infancy. In my opinion, this possibility should be taken into account in public health planning, and support should be given to more extensive and carefully planned studies.

Another interesting age group is school-children. As pointed out earlier, the content of iron in the brain continues to increase even after the age of 10 years. In two studies in Indonesia (Soemantri et al. 1985, and unpublished), there was a significant difference in school performance between iron-deficient (n=58) and non-iron-deficient children (n=72) around the age of 10 years. Moreover, giving iron orally (2 mg kg bw) for three months improved the school performance score significantly compared with children who were given a placebo. The trial was double-blind and the children were randomly allocated to the two groups. In children who were not iron-deficient, there was no effect of iron.

A similar study was made in Thailand (Pollitt et al. 1989) on 1358 9–11-year-old children. The language score was significantly lower in the iron-depleted children compared with the iron-replete. IQ and maths tests were also lower but not statistically significant. The effect of iron treatment (50 mg Fe/d for 2 weeks followed by 100 mg for 14 weeks) showed no significant effect. There was no overt difference in socio-economic status between the groups. However, radio and television was present more frequently in the homes of the iron-replete children. Differences in iron nutrition may thus not be the sole factor which explains differences in school performance. This study design did not include placebo groups.

Thyroid Hormone Metabolism and Thermoregulation (Dillman et al. 1980; Beard et al. 1984)

When subjects with iron deficiency anaemia are exposed to cold, the body temperature falls more than in controls. Women with iron deficiency anaemia showed signs of a poor thyroid hormone response, partly due to a reduced conversion of thyroxine to triiodothyronine (T_4 to T_3) in peripheral tissues. However, both T_4 and T_3 levels are reduced, suggesting a state of hypothyroidism. Another finding in iron-deficient women exposed to cold was increased levels of noradrenaline in plasma.

Much research is taking place in this area and there may be links both to enzyme deficiencies in peripheral tissue and to changes in the rate of secretion of thyroid-releasing hormone, in turn linked to changes in neurotransmitter metabolism. The practical importance of these interesting findings cannot yet be evaluated.

Effects on Immune Response (Dallman 1987; Vyas and Chandra 1984)

Iron plays a dual role in infectious diseases. It is essential for the growth of most bacteria, but it usually occurs in such chemical forms in the body that it is unavailable. On the other hand, iron is essential for some of the key immunological responses which may be impaired in a state of iron deficiency.

Several systems are involved in the defence against infections. The major ones are: The *cell-mediated* response with proliferation of T-lymphocytes; the *antibody-mediated* response with production of circulating antibodies by the B-lymphocytes: *phagocytosis* of bacteria and fungi by neutrophil leucocytes; the *complement system* that amplifies the immune response.

Iron is essential for the growth of T-lymphocytes. The reason is that DNA synthesis and cell growth are dependent on the enzyme ribonucleotide reductase, which requires a continuous supply of iron from transferrin (Thelander et al. 1983). There is convincing evidence that the cell-mediated immune response is impaired in iron deficiency.

The phagocytosis and killing of bacteria by neutrophils involves two steps: an uptake of the bacteria into the cells and their intracellular destruction. It is the latter step which is compromised in iron deficiency. Hydrogen peroxide and hydroxyl radicals – the oxidative burst – are responsible for the destruction and the enzymes responsible for this process contain iron. In clinical and experimental iron deficiency, there is a reduced intracellular destruction of bacteria, which is reversed 4–7 days after treatment with iron (Babior 1983).

Myeloperoxidase is an iron enzyme in the neutrophils that destroys certain microorganisms by another mechanism. The content of this enzyme is also reduced in iron deficiency.

The antibody-mediated immunity is not impaired in iron deficiency, probably because B-lymphocytes are less susceptible to a continuous supply of iron for their growth (Cramer et al. 1985).

There are several studies which suggest an association between iron deficiency and frequency of infection. However, the experimental design of most studies does not allow really valid conclusions (for a critical review, see Dallman 1987). Considering the important role of iron in the immune response, one would expect that future studies will demonstrate that iron deficiency is associated with increased morbidity in infections.

Concluding Comments

Iron has several key functions in the body. The maintenance of most of these vital functions is safeguarded by compensatory or substitutional mechanisms.

Recent animal studies strongly suggest that a long-lasting iron deficiency leads not only to a reduction in the haemoglobin level but also, at least, to a deprivation of tissue enzymes which impairs important functions in the body. This seems to be especially valid if the iron supply is insufficient during growth and development of the brain.

It is much more difficult to study the effects of iron deficiency in man. Results of studies on iron deficiency which has been induced by phlebotomy in short-term experiments probably bear little resemblance to the real situation of long-lasting iron deficiency. Moreover, it is difficult in epidemiological studies to isolate the effect of iron deficiency from the effects of several other factors which may influence physical and mental performance, well-being, rate of infections etc. The true effects of iron deficiency may thus be both underestimated and overestimated in human studies. Controlled studies in man on the effects of iron deficiency have only been carried out during the last decade and more has successively been learnt about bias factors influencing the interpretation of the results. More studies are needed but these must be very carefully planned and carried out to avoid the numerous bias factors involved in studies on more general symptoms, "well-being", learning etc.

Today, the balance of all evidence available in animal and human studies indicates that a deficiency of iron has negative effects on health and well-being. It is therefore reasonable that this position should be the basis for action in the public health field.

References

Ashkenazi R, Ben-Shackar D, Youdim MBH (1982) Nutritional iron and dopamin binding sites in the rat brain. Pharmacol Biochem Behav 17 [Suppl 1]:43–47

Babior BM (1983) The respiratory burst of phagocytes. J Clin Invest 73:599–601

Basta S (1974) Iron deficiency anaemia in adult males and work capacity. ScD Thesis, Massachusetts Institute of Technology, Cambridge, USA

Beard J, Green W, Miller L, Finch CA (1984) Effect of iron-deficiency anaemia on hormone levels and thermoregulation during cold exposure. Am J Physiol 247:R114–119

Cramer E, Pryzwansky KB, Villeval JL, Testa U, Breton-Gorius J (1985) Ultrastructural localization of lactoferrin and myeloperoxidase in human neutrophils by immunogold. Blood 65:423–432

Dallman PR (1986) Biochemical basis for the manifestations of iron deficiency. Am Rev Nutr 6:13–40

Dallman PR (1987) Iron deficiency and the immun response. Am J Clin Nutr 46:329–334

Dallman PR, Spirito RA (1977) Brain iron in the rat: extremely slow turnover in normal rats may explain long-lasting effects of early iron deficiency. J Nutr 107:1075–1081

Dallman PR, Siimes MA, Mauies EC (1975) Brain iron: persistent deficiency following short-term iron deprivation in the young rat. Br J Haematol 31:209–215.

Davies KJA, Donovan CM, Refino CJ, Brooks GA, Packer L, Dallman PR (1984) Distinguishing effects of anaemia and muscle iron deficiency on exercise bioenergetics in the rat. Am J Physiol 246:E535–543

Davies KJA, Maguire JJ, Brooks GA, Dallman PR, Packer L (1982) Muscle mitochondrial bioenergetics, oxygen supply, and work capacity during dietary iron deficiency and repletion. Am J Physiol 242:E418–427

Dillman E, Gale C, Green W, Johnson DG, Mackler B, Finch CA (1980) Hypothermia in iron deficiency due to altered triiodothyronine metabolism. Am J Physiol 239:R337–381

Edgerton VR, Bryant SL, Gillespie CA, Gardner GW (1972) Iron deficiency anaemia and physical performance and activity of rats. J Nutr 102:381–400

Edgerton VR, Ohira Y, Hettiaractichi J, Senewiratne B, Gardner GW, Barnard RJ (1981) Elevation of haemoglobin and work performance in iron deficiency subjects. J Nutr Sci Vitaminol 27:77–86

Finch CA, Miller LR, Inamdar AR, Person R, Seiler K, Mackler B (1976) Iron deficiency in the rat. Physiological and biochemical studies of muscle dysfunction. J Clin Invest 58:447–453

Gardner GW, Edgerton VR, Senewiratue B, Barnard RJ, Ohira Y (1977) Physical work capacity and metabolic stress in subjects with iron deficiency anaemia. Am J Clin Nutr 30:910–917

Glover J, Jacobs A (1972) Activity pattern of iron-deficient rats. Br Med J ii:627–628

Hallberg L (1982) Iron absorption and iron deficiency. Hum Nutr: Clin Nutr 36C:259–278

Hallberg L (1989) Search for nutritional confounding factors in the relationship between iron deficiency and brain function. Am J Clin Nutr 1989, 50: 598–606

Husaini, Karyadi D, Gunadi H (1983) Evaluation of nutritional anaemia intervention among anaemic female workers on a tea plantation. In: Hallberg L, Scrimshaw NS (eds). Iron deficiency and work performance. The Nutrition Foundation Washington DC, pp 73–78

Jefferies WA, Brandon MR, Hunt SV, Williams AF, Galter KC, Mason DY (1984) Transferrin receptor on endothelium of brain capillaries. Nature 312:162–163

Kaladhar M, Rao BSN (1982) Effects of iron deficiency on serotonin uptake in vitro by rat brain synaptic vesicles. J Neurochem 38:1576–1581

Lozoff B (1988) Behavioural alterations in iron deficiency. Adv Pediatr 35:331–360

Lozoff B, Brittenham GM (1986) Behavioural aspects of iron deficiency. Prog Hematol 14:23–53

Lozoff B, Brittenham GM, Viteri FE, Wolf AW, Urrutia JJ (1982a) Development deficits in iron-deficient infants: the effects of short-term oral iron therapy on developmental deficits in iron-deficient anaemic infants. J Pediatr 100:351–357

Lozoff B, Brittenham GM, Viteri FE, Wolf AW, Urrutia JJ (1982b) Developmental deficits in iron-deficient infants: effect of age and severity of iron lack. J Pediatr 101:948–952

Palti H, Devsner B, Adler B (1983) Does anaemia in infancy affect achievement on developmental and intelligence tests? Hum Biol 55:189–194

Pollitt E, Hathirat P, Kotchabhakdi NJ, Missell L, Valyasevi A (1989) Iron deficiency and educational achievement in Thailand. Am J Clin Nutr 1989, 50: 687–697

Pollitt E, Saco-Pollit C, Leibel RL, Viteri FE (1986) Iron deficiency and behavioural development in infants and preschool children. Am J Clin Nutr 43:555–565

Scrimshaw NS (1984) Functional consequences of iron deficiency in human populations. J Nutr Sci Vitaminol 30:47–63

Soemantri AG, Pollitt E, Kim I (1985) Iron deficiency anaemia and educational achievement. Am J Clin Nutr 42:1221–1228

Symes AL, Missala K, Sourkes TL (1971) Iron and riboflavin-dependent metabolism of a monoamine in the rat in vivo. Science 174:153–155

Thelander L, Gräslund A, Thelander M (1983) Continual presence of oxygen and iron required for mammalian ribonucleotide reduction: possible regulation mechanism. Biochem Biophys Res Commun 110:859–865

Viteri FE, Torun B (1974) Anaemia and physical work capacity. In: Garby L (ed) Clinics in haematology, vol 3, no 3. WB Saunders, Philadelphia pp 609–626

Voornees ML, Stuart MJ, Stockman JA, Oski FA (1975) Iron deficiency anaemia and increased urinary norepinephrin excretion. J Pediatr 86:542–547

Vyas D, Chandra RK (1984) Functional implications of iron deficiency. In: Stekel A (ed) Nutrition in infancy and childhood. Raven Press, New York, pp 45–59

Walter T, Kovalskys J, Stekel A (1983) Effect of mild iron deficiency on infant mental developmental scores. J Pediatr 102:519–522.

Weinberg J, Dallman PR, Levine S (1980) Iron deficiency during development in the rat: behavioural and physiological consequences. Pharmacol Biochem Behav 12:493–502

Weinberg J, Bert LP, Levine S, Dallman P (1981) Long-term effects of early iron deficiency on consummatory behaviour in the rat. Pharmacol Biochem Behav 14:447–453

Yehuda S, Youdim MBH (1984) The increased opiate action of B-endorphin in iron deficient rats: the possible involvement of dopamin. Eur J Pharmacol 104:245–251

Yehuda S, Youdim MBH, Mostofsky DI (1988) Brain iron deficiency causes reduced learning capacity in rats. Pharmacol Biochem Behav 25:141–144

Youdim MBH (ed) (1988) Brain iron: neurochemical and behavioural aspects. Taylor and Francis, London

Youdim MBH, Green AR (1977) Biogenic monoamine metabolism and functional activity in iron-deficient rats: behavioural correlates. In: Iron metabolism. CIBA Foundation Symposium, pp 201–223
Youdim MBH, Yehuda S (1985) Iron deficiency induces reversal of dopamin dependent circadian cycles: differential response to d-amphetamine and TRH. Peptides 6:851–855
Youdim MBH, Grahame-Smith D, Woods HF (1975) Some properties of human platelet monoamine oxidase in iron-deficiency anaemia. Clin Sci Mol Med 50:479–485

Chapter 15

Zinc: The Functional Significance of Marginal Deficiency

Brittmarie Sandström

The Functions of Zinc

The major biochemical function of zinc is as a constituent of metalloenzymes. The first described was carbonic anhydrase in 1940, and since then more than 200 different zinc enzymes have been identified in plant and animal tissue. Alcohol dehydrogenase, superoxide dismutase, DNA-polymerase, RNA-polymerase, alkaline phosphatase and carboxypeptidase are all zinc-metalloenzymes and examples can be found in each of the six major categories of enzymes. This means that zinc is involved in more or less every biochemical process in the body. In some of these enzymes zinc is present at the active site e.g., acting as an electron acceptor, in others and in non-enzyme proteins the function of zinc is structural as S–S bridges or cross-links between thiolates and imidazoles.

Zinc has important structural roles also at other sites. Bound zinc stabilizes the structures of RNA, DNA and ribosomes and zinc seems to have an important role in the structure and function of membranes (Bettger and O'Dell 1981).

There is a profound and rapid effect of zinc restriction on growth and development which seems to be due to impaired DNA synthesis rather than to protein synthesis. Recent studies suggest that zinc affects normal chromatin restructuring and gene expression (Chesters 1989).

Zinc plays an important role in immune defence. In the mouse, which has an immune system analogous to that of the human, a suboptimal zinc intake causes marked atrophy of the thymus, a reduction in leucocytes and in antibody-mediated, cell-mediated and delayed-type hypersensitivity responses (Fraker et al. 1987).

Indices of Zinc Status

Circulating levels of zinc in plasma or serum are the most often used indices of zinc status. Although plasma zinc is decreased in severe zinc deficiency the levels can be affected by a number of conditions that are unrelated to zinc status. Infection, fever, intake of a protein rich meal lowers plasma zinc, while long-term fasting, contaminated test-tubes and haemolysis increase the values found. Other biochemical indices used are zinc levels in white blood cells, erythrocytes and hair and urinary excretion of zinc. Of functional indices zinc tolerance test (Fickel et al. 1986), taste acuity (Bales et al. 1986) and dark adaptation (Sandström et al. 1987) have been tested. None of these biochemical and functional measures have so far been proven useful to identify marginal zinc deficiency in man. The reason for this could be the notable feature of zinc deficiency that overt symptoms can occur with almost no reduction in tissue concentration of zinc (Aggett et al. 1983), and that addition of zinc very rapidly improves the symptoms before any change in tissue content of zinc. Detection of mild zinc deficiency is, therefore, very difficult and at present no reliable, sensitive index of zinc status is available.

A promising approach is the development of a sensitive and rapid method to measure metallothionein. This sulphur-containing protein is induced by zinc, it binds zinc and seems to act as a storage form for zinc in tissues. In experimental animals the circulating levels of metallothionein in plasma and erythrocytes are correlated to tissue levels and to zinc status (Bremner et al. 1987). It is less sensitive than plasma zinc to infection and other trauma and might be useful for population surveys of zinc status.

Zinc in Food

Large variations in zinc content can be found between food sources that are otherwise nutritionally similar. Some examples of the relation between zinc and energy and protein contents are given in Table 15.1. In cereals, zinc is mainly located in the outer layer of the grain, and a low extraction rate of the grain means that the majority of the content of zinc as well as of other minerals is removed (Schroeder et al. 1967). In animal food sources, the fat content becomes one of the major determinants of the zinc content, as fat tissue has a much lower zinc content than muscle tissue. Green leafy vegetables and fruits are only modest sources of zinc.

Zinc Intake

Differences in zinc content of foods and specific food preferences can result in large differences in zinc intake, even within a population with similar intake of other nutrients. Fig. 15.1 shows the intake of zinc in relation to energy intake in a group of randomly selected Swedish women (Sandström 1982). Although there is an overall good correlation between energy and zinc intake, a number of subjects with a high energy intake had low zinc intakes and vice versa. The mean zinc intake was 9.1 ± 1.9 mg/d with a zinc density of 1.0 ± 0.2 mg/MJ. Similar figures have been reported in many dietary surveys. Sandstead et al. (1982) analysed data from large American nutrition surveys and concluded that energy intake was an important factor for zinc intake in the elderly. An estimated

Table 15.1. Zinc contents of selected foods expressed on a raw wet weight basis and in relation to their protein and energy contents (from data of Paul and Southgate 1978)

	mg/kg raw wet weight	mg/g protein	mg/MJ
Beef			
Lean	43	0.21	8.3
Fat	10	0.11	0.4
Pork			
Lean	24	0.12	3.9
Fat	4	0.06	0.1
Fish			
Cod	4	0.02	1.2
Milk	3.5	0.11	1.3
Butter	1.5	–	0.05
Lentils	31	0.13	2.4
Wheatflour			
Wholemeal	30	0.23	2.2
White	9	0.08	0.6
Rice (polished)	13	0.20	0.8
Potatoes	3	0.14	0.8

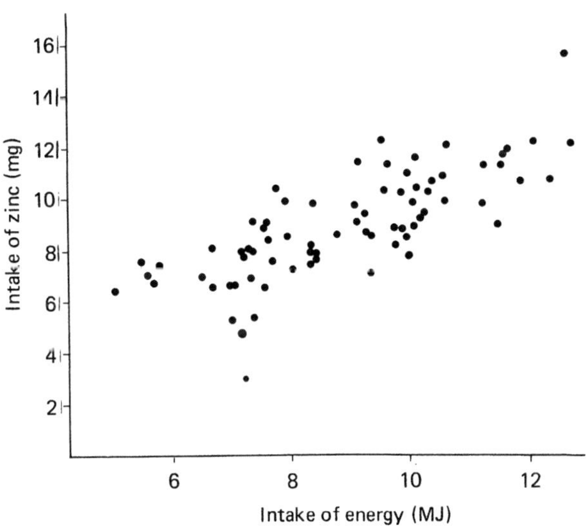

Fig. 15.1. Zinc intake in relation to energy intake in 72 Swedish women 44–45 years of age

zinc intake of 7.8 mg in the highest age group was the consequence of a low energy intake (5–6 MJ/day). The zinc density was 1.2–1.4 mg/MJ in this age group. Record et al. (1985) also observed a close correlation between energy and zinc intake in 73 pre-menopausal women completing a dietary frequency questionnaire. The mean zinc density was 1.5 mg/MJ, and women reporting energy intakes of 5 MJ/day had a zinc intake of only 6 mg/day.

The body's demand for zinc is increased during pregnancy. Reported zinc intakes of pregnant women are, however, of the same order as in non-pregnant women. A number of studies have shown an average intake of 9–12 mg/day and a zinc density of 1.1–1.4 mg/MJ (Hambidge et al. 1983; Hunt et al. 1983; Moser and Reynolds 1983; Abraham et al. 1985; Tuttle et al. 1985).

Interacting Dietary Factors

The studies cited above have not been able to show any correlation between zinc intake and proposed zinc status indices. This could be due to the insensitivity of the status indices, to poor methods for collecting zinc intake data and to the fact that dietary factors interacting with zinc absorption and utilization have not been accounted for. Organic substances acting as ligands facilitating uptake, or as complexing agents reducing solubility and absorption of zinc and other trace elements competing with zinc for uptake sites and carriers, can lead to a large variation in the fraction of total zinc intake that is absorbed. When a zinc salt is administered in a water solution 60%–80% is absorbed (Sandström et al. 1985, Sandström and Cederblad 1987), while the absorption from a mixed meal can vary from 5%–40% depending on its composition (Sandström 1989).

A dietary factor that could impair zinc absorption in man is phytic acid (myo-inositol hexaphosphate) in whole grain cereals, legumes and other vegetables. At pH values encountered in foods, phytic acid will be strongly negatively charged and will have a strong potential to bind positively charged molecules like zinc. The presence of phytic acid is the most likely reason for the low absorption of zinc observed from wholemeal bread and soy formula (Sandström et al. 1980; Nävert et al. 1985; Lönnerdal et al. 1984). Reduction of the phytic acid content of bread by long-term fermentation increases zinc absorption (Nävert et al. 1985). The absorption of zinc from a meal containing phytic acid can also be improved by an increased content of animal protein of the diet (Sandström et al. 1980; Sandström et al. 1989). Protein seems to act as an "anti-phytate" agent, which could be through a protein–phytic acid interaction not including zinc, or through an increased solubility and facilitation of zinc absorption by peptides and amino acids liberated during digestion. In the typical "modern" diet with a high degree of refined foods and a reasonable intake of animal protein, the phytic acid content is low and not likely to have significant effects on zinc absorption.

Elements with similar physicochemical properties may compete for binding sites and transport mechanisms. Such chemical similarity exists between zinc, cuprons and cadmium ions and interactions between these elements have been shown in model experiments. In the human diet the level of zinc is always higher than that of copper and cadmium, and in practice these elements are not likely to affect zinc absorption. In aqueous solutions of iron and zinc, iron depresses zinc absorption (Valberg et al. 1984, Sandström et al. 1985). Added to a meal, iron had no effect on zinc absorption indicating that iron enrichment, to the extent it is used in food, has no deleterious effects on zinc absorption.

A number of other nutrients or substances found in food have been shown in model experiments to affect zinc absorption (for review, see Sandström and Lönnerdal 1989). The nutritional significance of most of these substances when occurring in human diets is, however, uncertain.

Experimental Restriction of Zinc Intake in Man

Experimental dietary zinc restriction in normal adult subjects has shown that an intake of 0.17 mg zinc/day for 35 days gives a marked fall in tissue levels and urinary excretion of zinc (Hess et al. 1977), and signs of dermatitis and sore throats have also been observed in a zinc depletion study (Baer and King 1984). The role of zinc in host defence was also demonstrated by a decreased total lymphocyte count and impaired polymorphonuclear leucocyte chemotaxis (Baer et al. 1985).

Less severe restriction has little effect on indices of zinc status. Milne et al. (1987) fed 2.6 mg zinc for 125 days to five post-menopausal women without significant changes in plasma zinc, in the zinc content of blood cellular components or in the activity of zinc-containing enzymes in plasma or erythrocytes (alkaline phosphatase, carbonic anhydrase, amino laevulinic acid dehydratase, angiotensin-converting enzyme). Based on observations of decreased activity of liver alcohol dehydrogenase in zinc-deficient rats, they performed an ethanol tolerance test, and found a reduced ethanol clearance rate at the end of the low-zinc period which could be a functional change due to impaired zinc status. At higher intake levels, 5.5 mg/day for 8 weeks, neither urinary excretion or plasma levels changed significantly (Wada et al. 1985). However, decreased thyroid hormone levels, a lower basal metabolic rate, impaired protein utilization, elevated fasting blood glucose levels, decreased circulating levels of albumin, prealbumin, retinol-binding protein and transferrin were observed (Wada and King 1986).

Signs of Marginal Zinc Deficiency in Primates

Insights into the possible consequences of marginal zinc deficiency during pregnancy, infancy and early adolescence have been gained from long-term studies in rhesus monkeys. The level of zinc chosen for these experiments was such that plasma zinc concentrations did not change in non-pregnant females. Pregnant monkeys were given this otherwise nutritionally adequate diet during total pregnancy. Plasma zinc levels and plasma vitamin A were significantly lower by the third trimester, compared to pregnant control monkeys (Golub et al. 1984a, Baly et al. 1984). Immune functions seemed also to be affected by the marginal zinc intake demonstrated by a lower mitogen responsiveness of leucocytes (Golub et al. 1984a). The zinc-deprived group had more complications of pregnancy than the controls. Birthweight and length of male newborn monkeys were lower in the deficient group, while female newborns did not show any growth impairment (Golub et al. 1984b).

The male newborn monkeys with low birthweight also showed lower zinc and vitamin A levels. Another important observation was the significantly delayed skeletal maturation up to age 3 years, and radiographic evidence of defect bone mineralization up to age 6 months (Leek et al. 1988). A small group of zinc deprived male monkeys was also followed through adolescent growth spurt (Golub et al. 1988). Zinc-deficient animals had delayed onset of accelerated weight gain and linear growth, and the loss of subcutaneous fat typical for early adolescence did not occur. Behavioural performance was also affected in the zinc-deficient animals, and immune function was depressed.

Indications of Marginal Zinc Deficiency in Man

The observation in primates of a reduced growth rate at a marginal zinc intake has been confirmed in man by studies of controlled zinc supplementation in infants and children. Male, but not female, infants showed a significantly greater weight increment over the first 6 months with zinc supplementation compared to that of control infants (Walravens and Hambidge 1976). Similar results were observed for preschool children in low income families (Walravens et al. 1983). Supplementation was associated with an increase in energy and protein intake (Krebs et al. 1984).

Low levels of maternal serum zinc in pregnant women have been associated with prolonged labour, atonic bleeding, delivery outside normal term and congenital abnormalities (Jameson 1976). The nature of this association is, however, unclear. It has not been shown to be related to dietary intake of zinc, and zinc supplementation of 15–45 mg/day has so far failed to improve pregnancy outcome except for a possible reduction in the incidence of a dysfunctional labour pattern (for review see Swanson and King 1987).

Attempts to identify other signs of marginal zinc deficiency in otherwise "healthy" adult subjects has so far not been successful. Bogden et al. (1987) investigated 100 healthy elderly subjects and found a high incidence of anergy, and that the response to the tested antigens was correlated to plasma zinc levels of the subjects. In a second study 103 healthy elderly people were treated with 15 mg zinc/day, 100 mg zinc/day, or a placebo for 3 months (Bogden et al. 1988). Plasma zinc was significantly increased only in the group treated with 100 mg zinc, while zinc concentration in blood cells were not affected by any treatment. None of the treatments altered delayed dermal hypersensitivity or in vitro lymphocyte proliferation response to mitogens and antigens. The results suggest that impaired cellular immunity defence in the elderly is not in general a sign of marginal zinc deficiency.

Adaptation to Low Zinc Intake

The absence of signs of zinc deficiency even on low dietary zinc intakes can be explained by the body's efficient homeostatic mechanisms. Homeostasis is maintained by changes in absorption, gastrointestinal excretion and urinary and sweat losses (Milne et al. 1983; Jackson et al. 1984; Wada et al. 1985; Turnlund et al. 1986). In Table 15.2 examples are given of zinc utilization in young, healthy subjects at different zinc intakes. Changes in gastrointestinal excretion of zinc seem quantitatively of greater importance than changes in absorption for regulation of body content of zinc in these subjects. Jackson et al. (1988) have also studied zinc metabolism in five undernourished, lactating women in Manaus, Brazil. Their normal diet, consisting of cassava flour, white bread and fish, provided only 6–11 mg zinc/day. A high fractional absorption of zinc (59%–84%) was observed, and in eight of ten balance periods a positive balance was calculated. These data suggest that it is possible to adapt to a chronically low intake of zinc. Whether all subjects possess this ability, and at what level of intake the risk for functional disturbances is increased, are questions that remain to be answered.

Table 15.2. Zinc utilization at normal and low zinc intake (mg/day). Data from Jackson et al. (1984) and Wada et al. (1985)

Intake	Absorption	Urinary zinc	Endogenous faecal zinc
15	4.7 (32%)	0.6	4.6
7	3.3 (47%)	0.6	3.0
5.5	2.7 (49%)	0.6	1.7

Conclusions

Modern lifestyles, including a low energy intake and – of probably greater importance – an increasing intake of foods rich in fat and sugar with a low zinc density, can give a low intake of zinc. The results from the limited number of experimental zinc depletion studies in man indicate that an intake of 5.5 mg zinc/day can give functional disturbances of energy metabolism. With a reasonable safety margin a daily zinc intake below 8 mg could pose a problem. This corresponds to an energy intake of 7–8 MJ with present dietary habits. Many women, constant dieters and many elderly people have a lower energy intake and must be considered at risk of inadequate zinc intake.

The only sign of mild zinc deficiency that clearly has been identified in man is a reduced growth rate in infants. However, the extent of marginal zinc status in infants as well as in the total population cannot be evaluated before sensitive functional indices or laboratory methods are available, and have been validated by adequately designed, randomized, controlled studies of dietary supplementation with physiological quantities of zinc. More knowledge is also needed about zinc metabolism at different intake levels and during periods of an increased requirement of zinc such as infancy, pregnancy and lactation; about the ability to adapt to a low zinc intake and about the utilization of zinc from different diets.

References

Abraham R, Campbell-Brown M, Haines AP, North WRS, Hainsworth V, McFadyen IR (1985) Diet during pregnancy in an Asian community in Britain – energy, protein, zinc, copper, fibre and calcium. Human Nutr Appl Nutr 39A:23–35

Aggett PJ, Crofton RW, Chapman M, Humphries WR, Mills CF (1983) Plasma leucocyte and tissue zinc concentrations in young zinc deficient pigs. Pediatr Res 17:433–445

Baer MJ, King JC (1984) Tissue zinc levels and zinc excretion during experimental zinc depletion in young men. Am J Clin Nutr 39:556–570

Baer MT, King JC, Tamura T et al. (1985) Nitrogen utilization, enzyme activity, glucose intolerance and leucocyte chemotaxis in human experimental zinc depletion. Am J Clin Nutr 41:1220–1235

Bales CW, Steinman LC, Freeland-Graves JH, Stone JM, Young RK (1986) The effect of age on plasma zinc uptake and taste acuity. Am J Clin Nutr 44:664–669

Baly DL, Golub MS, Gershwin ME, Hurley LS (1984) Studies on marginal zinc deprivation in rhesus monkeys. III. Effects on vitamin A metabolism. Am J Clin Nutr 40:119–207

Bettger WJ, O'Dell BL (1981) A critical physiological role of zinc in the structure and function of biomembranes. Life Sci 28:1425–1438

Bogden JD, Oleske JM, Munves EM et al. (1987) Zinc and immunocompentence in the elderly: baseline data on zinc nutriture and immunity in unsupplemented subjects. Am J Clin Nutr 45:101–109

Bogden JD, Oleske JM, Lavenhar MA et al. (1988) Zinc and immunocompentence in elderly people: effects of zinc supplementation for 3 months. Am J Clin Nutr 48:655–663

Bremner I, Morrison JN, Wood AM, Arthur JR (1987) Effects of changes in dietary zinc, copper and selenium supply and of endotoxin administration on metallothionein I concentrations in blood cells and urine in the rat. J Nutr 117:1595–1602

Chesters JK (1989) Biochemistry of zinc in cell division and tissue growth. In: Mills CF (ed) Zinc in human biology. Springer-Verlag, Berlin, pp 109–118

Fickel JJ, Freeland-Graves JH, Roby MJ (1986) Zinc tolerance tests in zinc deficient and zinc supplemented diets. Am J Clin Nutr 43:47–58

Fraker PJ, Jardieu P, Cook J (1987) Zinc deficiency and immune function. Arch Dermatol 123:1699–1701

Golub MS, Gershwin ME, Hurley LS, Baly DL, Hendrickx AG (1984a) Studies of marginal zinc deprivation in rhesus monkeys. I. Influence on pregnant dams. Am J Clin Nutr 39:265–280

Golub MS, Gershwin ME, Hurley LS, Baly DL, Hendrickx AG (1984b) Studies of marginal zinc deprivation in rhesus monkeys. II. Pregnancy outcome. Am J Clin Nutr 39:879–887

Golub MS, Gershwin ME, Hurley LS, Hendrickx AG (1988) Studies of marginal zinc deprivation in rhesus monkeys. VIII. Effects in early adolescence. Am J Clin Nutr 47:1046–1051

Hambidge KM, Krebs NF, Jacobs MA, Favier A, Guyette L, Ikle DN (1983) Zinc nutritional status during pregnancy: a longitudinal study. Am J Clin Nutr 37:429–442

Hess FM, King JC, Margen S (1977) Zinc excretion in young women on low zinc intakes and oral contraceptive agents. J Nutr 107:1610–1620

Hunt IF, Murphy NJ, Cleaver AE et al. (1983) Zinc supplementation during pregnancy: zinc concentration of serum and hair from low-income women of Mexican descent. Am Clin Nutr 37:572–582

Jackson MJ, Jones DA, Edwards RHT, Swainbank IG, Coleman ML (1984) Zinc homeostasis in man: studies using a new stable isotope-dilution technique. Br J Nutr 51:199–208

Jackson MJ, Giugliano R, Giugliano LG, Oliveira EF, Shrimpton R, Swainbank IG (1988) Stable isotope metabolic studies of zinc nutrition in slum-dwelling lactating women in the Amazon valley. Br J Nutr 59:193–203

Jameson S (1976) Effects of zinc deficiency in human reproduction. Acta Med Scand 593 [Suppl]:1–89

Krebs F, Hambidge KM, Walravens PA (1984) Increased food intake of young children receiving a zinc supplement. Am J Dis Child 138:270–273

Leek JC, Keen CL, Vogler JB et al. (1988) Long-term marginal zinc deprivation in rhesus monkeys. IV. Effects on skeletal growth and mineralization. Am J Clin Nutr 47:889–895

Lönnerdal B, Cederblad Å, Davidsson L, Sandström B (1984) The effect of individual components of soy formula and cow's milk formula on zinc bioavailability. Am J Clin Nutr 40:1064–1070

Milne DB, Canfield WK, Mahalko JR, Sandstead HH (1983) Effect of dietary zinc on whole body surface loss of zinc: impact on estimation of zinc retention by balance method. Am J Clin Nutr 38:181–186

Milne DB, Canfield WK, Gallagher SK, Hunt JR, Klevay LM (1987) Ethanol metabolism in postmenopausal women fed a diet marginal in zinc. Am J Clin Nutr 46:688–693

Moser PB, Reynolds RD (1983) Dietary zinc intake and zinc concentrations of plasma, erythrocytes, and breast milk in antepartum and postpartum lactating and nonlactating women: a longitudinal study. Am J Clin Nutr 38:101–108

Nävert B, Sandström B, Cederblad Å (1985) Reduction of the phytate content of bran by leavening in bread and its effect on absorption of zinc in man. Br J Nutr 53:47–53

Paul AA, Southgate DAT (1978) McCance and Widdowson's The composition of foods, 4th edn. Elsevier/North-Holland, Biomedical Press, Amsterdam, New York, Oxford
Record IR, Record SJ, Dreosti IE, Rohan TE (1985) Dietary zinc intake of pre-menopausal women. Human Nutr Appl Nutr 39A:363–369
Sandstead HH, Henriksen LK, Greger JL, Prasad AS, Good RA (1982) Zinc nutriture in the elderly in relation to taste acuity, immune response, and wound healing. Am J Clin Nutr 36:1046–1059
Sandström B (1982) Zinc intake in Sweden. Näringsforskning 26:163–164
Sandström B (1989) Dietary pattern and zinc supply. In: Mills CF (ed) Zinc in human biology. Springer-Verlag, Berlin, pp 351–363
Sandström B, Cederblad Å (1987) Effect of ascorbic acid on the absorption of zinc and calcium in man. Int J Vit Nutr Res 57:87–90
Sandström B, Lönnerdal B (1989) Promoters and antagonists of zinc absorption. In: Mills CF (ed) Zinc in human biology. Springer-Verlag, Berlin, pp 57–78
Sandström B, Arvidsson B, Cederblad Å, Björn-Rasmussen E (1980) Zinc absorption from composite meals. I. The significance of wheat extraction rate, zinc, calcium and protein content in meals based on bread. Am J Clin Nutr 33:739–745
Sandström B, Davidsson L, Cederblad Å, Lönnerdal B (1985) Oral iron, dietary ligands and zinc absorption. J Nutr 115:411–414
Sandström B, Davidsson L, Lundell L, Olbe L (1987) Zinc status and dark adaptation in patients subjected to total gastrectomy: Effect of zinc supplementation. Human Nutr Clin Nutr 41C:235–242
Sandström B, Almgren A, Kivistö B, Cederblad Å (1989) Effect of protein level and protein source on zinc absorption in humans. J Nutr 19:48–53
Schroeder HA, Nason AP, Tipton IH, Balassa JJ (1967) Essential trace metals in man: Zinc. Relation to environmental cadmium. J Chron Dis 20:079–210
Swanson CA, King JC (1987) Zinc and pregnancy outcome. Am J Clin Nutr 46:763–771
Turnlund JR, Durkin N, Costa F, Margen S (1986) Stable isotope studies of zinc absorption and retention in young and elderly men. J Nutr 116:1239–1247
Tuttle S, Aggett PJ, Campbell D, MacGillivray I (1985) Zinc and copper nutrition in human pregnancy: a longitudinal study in normal primigravidae and in primigravidae at risk of delivering a growth retarded baby. Am J Clin Nutr 41:1032–1041
Valberg LS, Flanagan PR, Chamberlain MJ (1984) Effects of iron, tin, and copper on zinc absorption in humans. Am J Clin Nutr 40:536–541
Wada L, Turnlund JR, King JC (1985) Zinc utilization in young men fed adequate and low zinc intakes. J Nutr 115:1345–1354
Wada L, King C (1986) Effect of low zinc intakes on basal metabolic rate, thyroid hormones and protein utilization in adult men. J Nutr 116:1045–1053
Walravens PA, Hambidge KM (1976) Growth of infants fed a zinc supplemented formula. Am J Clin Nutr 29:1114–1121
Walravens PA, Krebs NF, Hambidge KM (1983) Linear growth of low income preschool children receiving a zinc supplement. Am J Clin Nutr 38:195–201

Chapter 16

Lowered Dietary Energy Consumption and Potential Consequences for Micronutrient Intake: An Overview

R.G. Whitehead

Introduction

In some ways my task of summing-up this volume is an easy one – the papers that have been contributed are sensible and to the point. I do not have to try and engineer sense out of nonsense. The honesty of the contributors does mean, however, that we are left with more unsolved problems than solved ones. Clearly what I must try to do, is to rationalize these and pull out some basic principles which need to be considered when we are planning future strategies. These will be both *research protocols* as well as plans for practical *public health programmes*. This overview can represent only my own personal opinions: the more consensus type of evaluation has been performed during the preparation of the Reports of the Vitamins and Minerals Working Groups.

The title given to this volume implies that there have been major changes in lifestyle during recent years and that these have implications for both dietary energy and micronutrient intake. The assumption made by most speakers is that a more sedentary lifestyle, coupled with a fashionable desire to remain young and especially slim throughout one's life-span has led to people consuming much less food than they used to and this has had a "knock on" effect on mineral and vitamin intake.

Has There Really Been a Change in Total Food Intake?

A number of contributors have produced evidence which indicates that the total energy intake of many people in Europe is now less than national and international RDA values but, as we have observed, if we are to be sure there has been a *change*, we must examine *sequential* data spanning a number of years.

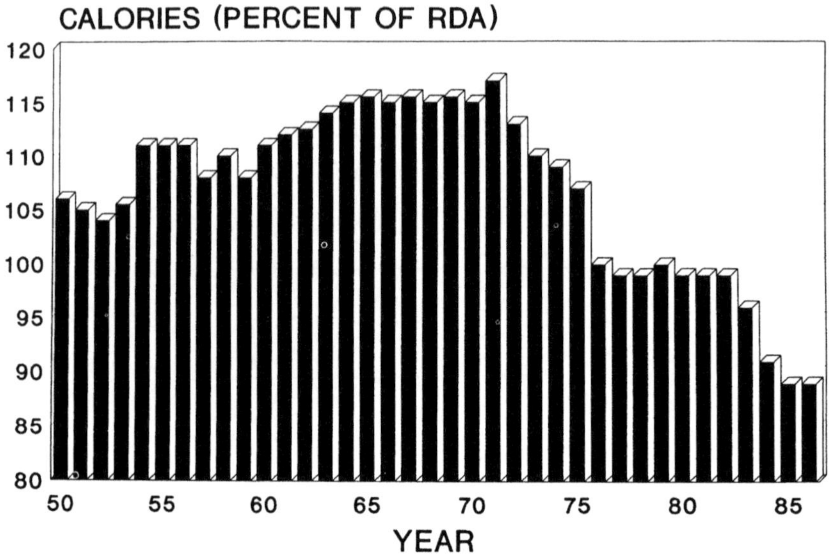

Fig. 16.1. Changes in the energy intakes of British families between 1948 and 1986 relative to the recommended dietary allowance (RDA): calculated from various Reports of the British Nutritional Food Survey (Ministry of Agriculture, Fisheries and Food).

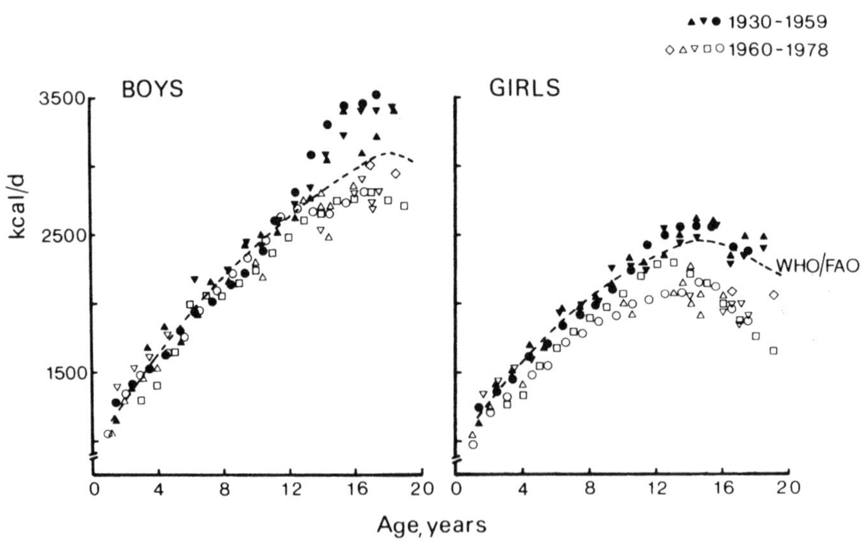

Fig. 16.2. Energy intakes (kcal/d) of boys and girls studied between 1930 and 1955, compared with more recent studies, in relation to WHO/FAO (1973) recommendations. For details of studies cited see Whitehead, Paul and Cole (1982) Human nutrition. Appl Nutr 36:57–62.

At the risk of being accused insular I would like to present data from the British National Food Survey covering the period 1948–1986: after a period of relative gluttony following the Second World War there was a plateau in energy intake during the 1960s followed by a mathematically well-defined fall, starting around 1970 (Fig. 16.1). For the UK there is also substantial information that this lowered energy intake is spread right throughout the various age groups. Fig. 16.2, for example, illustrates the change in energy intakes of British children aged 0–20 years: the gap during puberty is especially dramatic.

A number of contributors to this volume have paid special attention to pregnancy and lactation, at which times in the reproductive life of a woman we have always believed there needs to be a substantial increase in both dietary energy and nutrient intake. The tacit assumption is that if one consumes more calories one will automatically correct any potential micronutrient deficit. But this increased food intake no longer seems to be common practice. The Nestlé Foundation funded an international collaborative investigation of energy homeostatis during pregnancy and lactation and in all our participating countries we failed to find energy increments anything like the theoretically recommended one.

Are Data Devised from Food Intake Measurements of Misleading Accuracy?

Can we be sure that the above data are a true representation of what is really happening? The problem of misleading food composition tables, on which most assessments of energy and nutrient intake are based, has been raised a number of times in this volume but we must also bear in mind that the quantification of food *consumption* is also full of methodological pitfalls. These are not so much errors of measurement but the production of atypical data arising from the inevitable constraints imposed by investigation.

In the past we have had little or no opportunity to test the validity of our assumptions about intakes necessary for energy homeostasis in free-living persons. Fortunately this is now being changed by the development of very sensitive mass spectrometers which can measure changes over time in ^{2}H and ^{18}O enrichment of the urine and saliva of people dosed with these stable isotopes. From such measurements an accurate mathematical evaluation of energy expenditure can be obtained. If one also has an accurate measure of basal metabolic rate (BMR) then one can calculate the amount of energy expended on activity (plus the small component arising from diet-induced thermogenesis).

My colleagues in Cambridge have now applied this approach to a wide range of lifestyles and age groups. Since there has quite rightly been so much emphasis on women in this volume, let me present just one example from our data on Cambridge women (Fig. 16.3). The typical woman spends only an average of 550 kcal/d on all daily activities above that which would be required if she were asleep for the whole of the day. This is only 1.38×BMR and for some individuals the factor is even less. At one time we used to consider 1.5×BMR the minimum energy requirement for meaningful survival: for a woman a factor of 1.5 would now represent an above-average rate of activity!

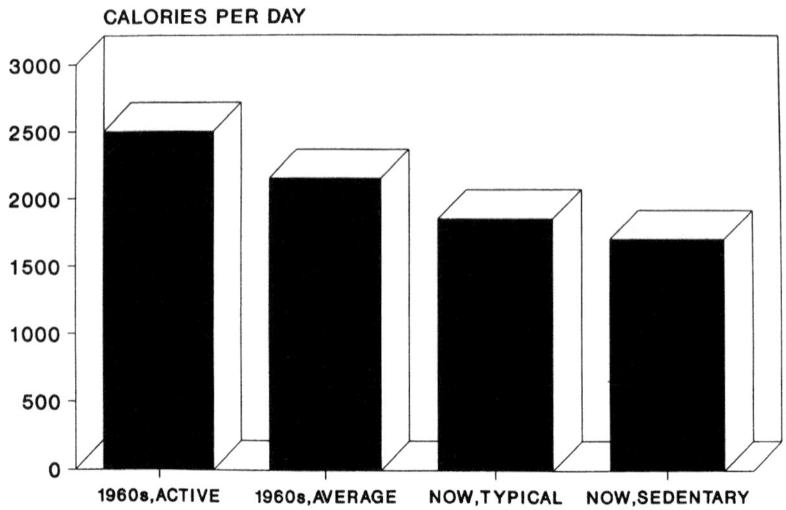

Fig. 16.3. Changes in energy expenditure of British women between the 1960s and the 1980s. The more modern data have been confirmed using the doubly-labelled water technique.

These data are by no means atypical. I could present similar findings from children and older people: a sedentary lifestyle involving a low energy expenditure seems omnipresent.

I have gone into this newly developed methodology, not just to fill in an important gap in our deliberations, but to emphasize a key investigative principle for the future. In the prospective cohort studies advocated in this volume for the better characterization of secular trends, we must no longer guess at what is happening vis-à-vis activities amongst different population groupings. Nutrition has far too long been a semi-quantitative science, almost a folk art: we must take full advantage of modern technological advances in order to produce data of sufficient accuracy for rational public health planning.

Can One Automatically Link Energy Intake to a Reduced Micronutrient Intake?

To summarize this overview so far: I think we can be reasonably confident that an increasingly sedentary lifestyle, coupled with a cosmetic desire to be slim, *is* resulting in lower energy intakes than we used to have. Is this mirrored, exactly, by a similar nutrient intake reduction? This must obviously depend on precisely what people have left out of their diet and how the farmer and the food industry have responded to changes in demand by altering the composition of their products.

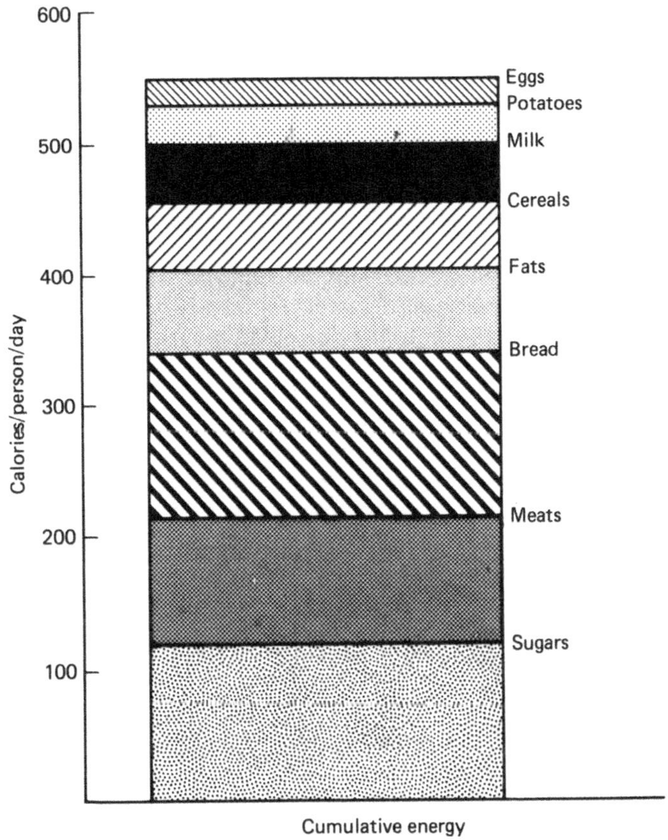

Fig. 16.4. Specific types of food affected by the reduction in average family energy intake between 1970 and 1986 (see Fig. 16.1).

Ann Prentice has discussed the changes in calcium containing foods in the UK (Chap. 12). Let us look at the picture insofar as energy is concerned. Fig. 16.4 demonstrates that in the UK the biggest energy contributors to the 533 kcal energy drop that has occurred since 1970 are sugar and preserves followed by meat and bread. Vegetables and fruit have been virtually unaffected. Thus our consumption of many water-soluble vitamins has shown very little fall and changes in calcium intake from milk have exhibited only a moderate drop and this has been partially corrected by an increased consumption of cheese. In other words, the *overall density* of many nutrients in the British diet has risen.

The fall in red meat consumption is a concern for us as meat is an important vehicle for iron intake as well as for other divalent minerals. However, as a number of contributors have indicated, the move towards leaner meats, the judicious butchering of meat and modifications to cooking has produced a partial compensatory effect.

Practical Implications for the Immediate Future

What does this mean in practical terms? At an analytical level we clearly need closer surveillance of food composition so that we know more precisely what is happening to our food. Nutritionists also need to collaborate more closely with the food industry in order to ensure a rational nutrient density in their products. This is especially so in the case of "whole meal" convenience foods and for chains of "fast food" restaurants.

There can be no "blank cheque" for nutrient supplementation, however. Much can be achieved by the careful selection of different food constituents. I would not deny that in certain circumstances there may be a call for supplementation, especially where diets for *special population groups* are concerned, but this will be the exception rather than the rule.

The recognition that a reduced energy expenditure coupled with a lowered physiological need for dietary energy can have health sequelae, in terms other than, just for obesity or slimness, needs to be taken more into account by health educators as well. The days have gone when we could rely on the *bulk* of food we needed to consume to satisfy our appetites being able to compensate for nutritionally poor quality diets. Dietary excellence is now just as much an issue for the average housewife as for the "bon vivant" and the "gourmet".

Low Fat, Low Sugar Food Products

We must also stop considering the fat and sucrose content of our diet just in terms of the pathogenesis of cardiovascular disease. We now have a much more basic reason for limiting the dietary consumption of these components. There is no way that we can equate present-day energy requirements with correspondingly low intakes, if fats and sucrose are too prevalent in our food. Moderation and care in menu design is also essential if it is to have the desirable nutrient density.

Lowered Energy Intakes and RDAs for the Micronutrients

Virtually all the contributors have addressed the question as to whether our present nutrient and energy intakes are compatible with recommended dietary allowances. This, of course, depends on which RDAs you adopt as the reference. Generally the Americans have adopted the highest values and WHO/FAO have shown more moderation. Many countries on the continent of Europe have inclined to the American philosophy while the British conclusions tend to be closer to those of the UN agencies. Politics, national interests and fashion have led as much to these disparate conclusions as science. Indeed, science has often been able to make very little contribution because of a paucity of truly relevant information.

Why should this be so? Although the derivation of RDAs is not the primary concern of this volume, contributors have inevitably been dragged into a consideration of RDAs otherwise they could not interpret the significance of relatively low energy intakes! Professor Brubacher has summarized the various approaches currently being adopted in order to develop more rational RDAs: (a) the analytic approach, (b) the logical approach (which I prefer to call the *physiological* approach), (c) the *epidemiological* approach, (d) the *balance* approach and (e) the *pragmatic* one.

If we are honest with ourselves it is mostly this last approach that we have been forced to fall back upon. If a population has a certain range of intakes and is seemingly healthy the pragmatic assumption is that there cannot be greater need for that nutrient.

Critical contributors have, however, pointed out that this approach is dependent on the appearance of rather gross abnormalities and what we should be concentrating upon is optimum *health* rather than *disease*.

The physiological and the epidemiological approaches are more health orientated but they need some *functional* parameter against which intake can be compared. For most nutrients, however, such functional parameters do not exist. Before we can justify detailed, intensive and costly investigations of these types in the metabolic ward or in the community, more fundamental research into the pathogenisis of nutrient disorders needs to be carried out. Only in this way are we going to be able to identify the critical biochemical, physiological or behavioural link which we can use as the basis for that functional parameter. It is embarrassing for a scientist like myself, who has sat on a number of "expert" committees over the past decades, to see so little new progress in this crucial area.

The development of relevant measurement parameters for nutrient function for subsequent physiological and epidemiological study must be one of our major priorities. It is in the interest of the food industries and public health authorities alike to clear this log-jam. As I have already indicated, we biologists must learn to take better advantage of technological developments so that we can study more exactly the physiology of people under "free-living" conditions. We have done it with energy but similar opportunities are now available in the realms of vitamin and mineral homeostasis and the funds need to be found, so that these opportunities can be exploited.

The Complication of Bio-availability

Bio-availability of nutrients within the diet has been another complication for participants. This is a complex issue and it is always going to be a problem area. Within the context of this meeting we need to keep in mind that current dietary trends are not only affecting the basic composition of the diet we consume but also the bio-availability of constituent micronutrients. When we encourage people to eat more "dietary fibre", invariably they consume more phytate too with the concommitant risk of chelating a greater proportion of divalent ions such as Ca^{2+}, Fe^{2+}, Zn^{2+} and Mg^{2+}.

Do we actually need to allow for this? There is good evidence that man can adapt to lower than customary intakes of the different nutrients – as Dr Sandstroem has so succinctly put it – the more you eat, the more you need. She was specifically referring to zinc but it could so easily have been calcium, iron or magnesium.

Metabolic Interactions between Nutrients

Nutrient interactions are also something we should be paying more adequate attention to. Professor Hallberg has pointed out that vitamin C has a newly recognised role to play in the diet, insofar as iron needs are concerned: because of our greater consumption of "dietary fibre" and the phytates, an increased vitamin C consumption would convert more of the Fe^{2+} to Fe^{3+} thus reducing the chelation effect and making more of the iron available to cover physiological needs.

Interaction complexities are almost limitless in their potential importance. Calcium provides another good example. Absorption and skeletal calcium balance are both dependent on vitamin D status. This is *primarily* determined by sunlight exposure which can vary widely within Europe. Rationalizing the practical ramifications of these variations is far from simple, however. When it is cold and wet, as in the UK, people stay indoors: but, likewise, when it is very hot and sunny they often do the same. Furthermore, the attractions of television are not only encouraging a low energy expenditure, they are also contributing to a reduced sunlight exposure and hence, potentially, adversely affecting vitamin D conversion as well.

The Diagnosis of Community Nutritional Status from Intake Data

An important outcome of these academic considerations is the danger of over simplified conclusions. It is virtually impossible to diagnose the likelihood of potential clinical deficiency from RDAs alone. Not only are there substantial inter-individual physiological variations in requirement, bio-availability, nutrient interactions and lifestyles all complicate the interpretation of intake data. An intake which might appear small could be quite adequate under a particular set of environmental circumstances and, of course, vice-versa. The only way to be sure of whether or not there is a potential problem is via the actual measurement of some biological parameter in the suspect population. And this brings us back to the need for those elusive "functional tests"!

Conclusions

In my opinion we cannot, at this point in time, say with any degree of certainty that the lowered energy intakes resulting from an inactive lifestyle are resulting in positive harm in terms of the micronutrients. Without doubt there will be minorities involved in over-intensive slimming on ill-considered diets who have a problem but for people in energy balance the situation is not so clear cut. Obviously one can demonstrate intakes below national and international RDAs but the latter do have built-in safety margins and they do not make allowances for physiological adaptation when total intake is low.

I am not recommending complacency however. Dietary planning needs even more stringent consideration now than ever before. Regular public health surveillance must be stepped up. Whilst we may not be able to interpret with absolute confidence "single shot" surveys, prospectively conducted programmes in which intakes or plasma biochemical values are measured at regular intervals can be very informative. We need to know how situations are developing and whether or not this is potentially good or bad news in health terms.

Health educators need to be given more exact information on the significance of current energy intakes: the public should be advised about the need for greater dietary care in the planning of the family menu now that energy intakes are lower than they were. Likewise restauranteurs and the manufacturers of convenience foods need to recognize that many of their clients will welcome the possibility of choosing menus low in dietary energy (low fat, low sucrose) but relatively high in micronutrient density. The nutrient content and the balance between different nutrients can no longer be taken for granted. Wisdom would dictate greater care and attention to nutrient density especially insofar as micronutrients are concerned.

Appendix 1

Report of the Vitamins Working Group

Chairman: K. Pietrzik
Rapporteur: A. Kallner
Members: O. Amédée-Manesme, R. Buzina, F. Fidanza, K. F. Gey, L. Kohlmeier, J. Schrijver, H. van den Berg
Observers: P. Bay, R. Hawkins, D. Hornig, J. M. M. Van Amerlsvoort

The working group initially discussed what should be understood by a modern lifestyle. Although this was one of the major subjects presented before the beginning of the workshop it appeared necessary to compare the experience and expectations of different countries and social strata. It was generally agreed that there is a considerable difference between countries and that the term "modern lifestyle" is a continuous process and a direction of development which eventually the majority of the population will follow. It is likely that changes will occur sooner and more homogeneously in urban than in rural populations.

"Modern lifestyle" has general implications which go far beyond nutrition. Some major items of importance in evaluating the impact of lifestyle on vitamin status were identified (Table A1.1).

We do not know enough of the quantitative aspects of the various aspects of modern lifestyle: we do not know if morbidity or mortality has changed due to changes in the lifestyle.

Table A1.1. Modern lifestyle

Increased urban population
Smaller families
Increasing number of elderly
Increasing information and communication
Increased food from external catering
Less home cooking
Increasing variety of refined foods
Changing types and availability of raw materials
Influence of modern nutritional advice
Changing energy balance
Modern weight ideals
Changing physical activity
Changing drinking habits
Changing smoking habits
Changing medication

There is growing concern about the effects of environmental and dietary factors on health. Therefore one might expect that more people will follow the present recommendations on intakes of less energy, less fat and more fibre. Very few hard data are available on the consequences of marginal deficiencies of micronutrients, particularly the vitamins.

Where there is increasing intake of foods of high energy density or “empty calories” greater emphasis will have to be laid on the intake of other foods with a high nutrient density. As regards beverage consumption, it was pointed out that some are supplemented with several vitamins, whereas others may be completely devoid of vitamins. Special attention should be paid to alcoholic drinks, particularly because of an increased demand for vitamin B1.

The food industry is not only concerned with the production of refined foods but also with changing and improving raw materials by conventional selection of cattle and crops, aquaculture and methods of feeding livestock, together with new developments in biotechnology. The degree to which these procedures might change the composition of food is not known.

The introduction of more refined, semi-prepared food and out-of-home services was regarded as a potential improvement of nutrition. Strict regulations and consciousness among manufacturers and consumers have resulted in a better control of food additives. This is an ongoing process where further achievements are expected. However, further improvement of these types of food requires that they are properly formulated with regard to a sound balance of micronutrients and energy. It is important that agreement is reached on the quality assessment of refined foods and how their contents should be declared. The concept of nutrient density was raised as a useful index quality.

Labelling of pre-or semi-fabricated foods is thus of great importance but requires education of the consumers to make sure that the information provided is understood and is useful. Labelling can, in itself, be educational. A quantitative labelling system could also be envisaged by which certain types of foodstuffs with a defined nutrient density would be understood by the less well-informed consumer.

In order to formulate fields of particular priority for research the vitamins were discussed individually.

It was concluded that micronutrient deficiencies do occur in modern European communities. These, however, rarely result in overt dificiency symptoms. While food composition tables are frequently used to evaluate diets it was recognized that they are often incorrect e.g., with respect to vitamin E contents.

It was concluded that for vitamins A, E, C, D, folate and riboflavin useful clinical or biochemical indicators of vitamin status are available. However, functional tests and their links with blood levels are often lacking. Some are promising; for instance, the impression cytology test is available for assessment of vitamin A status, the pentane breath test seems to be related to the lipid standardized vitamin E level in serum provided the vitamin C and selenium status are adequate. Standardization of plasma folate is difficult but the hypersegmentation of granulocytes could be used to make results compatible between laboratories, research groups and other users of the data.

The health implications of a decreased vitamin status are well-documented. Major health problems in Europe are cardiovascular diseases and cancer. Epidemiological studies have indicated a link between vitamin status (i.e., vitamin A, vitamin E, ß-carotene and vitamin C) and these conditions. In certain

Table A1.2. Education topics

Food composition and contents
Nutritional needs
Balance of diets
Understanding of labelling
Chemistry of cooking
Population information with special reference to risk groups

population groups there are special problems, e.g., osteomalacia in institutionalized and/or immobilized individuals, as well as among certain groups of immigrants which may be linked to an impaired vitamin D status; pregnancy may predispose folate deficiency; multifunctional deficiencies of vitamins can be expected in the elderly; and riboflavin deficiency in countries with low milk consumption. Recent reports indicate that considerable proportions of certain population groups have plasma-ascorbate levels which indicate an intake below optimal.

Many of the problems that have been outlined during the discussions could probably be overcome or diminished by education of the population as well as of certain key professionals. Some particular areas where these educational efforts should be focused were identified (Table A1.2). It was realized that in order to achieve this all possible media should be used, e.g., news magazines and TV but also schools and other means to influence public opinion. The need for increased education is considerable and long-term. The responsibilities of governments and industry were discussed. It was concluded that industry should be obliged to contribute but that governments and their educational systems have the ultimate responsibility for continuous and comprehensive education.

Table A1.3. Research and development projects

1. *Lifestyle*
   Methodology to estimate energy balance
   Monitor energy intake and the change of micronutrient density in longitudinal studies
   What do people eat – Temporal considerations with a projection to the future
   European surveillance of nutritional status with regard to micronutrients
   Simulation studies on consumption models
2. *Food consumption and composition*
   Micronutrient assay methodology
   Harmonization of quality assessment
   Quality assessment of refined food
   Consumption modelling of energy intake and its influence on nutrient density of vitamins
   Creation of a data base on food intake
3. *Nutrition status*
   Definitions of terminology
   Data interpretation
   Reference intervals
   Idenitification of risk groups with regard to marginal deficiencies of vitamins and minerals
   Functional tests for vitamins C and B6 and riboflavin
   More data on vitamin A (fractions), ß-carotene and other carotenoids
4. *Health impact*
   Vitamin status and chronic diseases
   Interactions between different vitamins and other food components
   Drug interactions with vitamins

The discussions were summarized by formulating a number of research and development projects which address each of these subjects (Table A1.3).

Based on the deliberations of the group the topic was subdivided into four categories:

1. Lifestyle
2. Food consumption
3. Nutritional status
4. Health impact

This brief catalogue of projects indicates areas where projects should be initiated and developed. Each of the categories may be further elaborated and expanded. The working group did not discuss the list in order of priority and some subjects certainly have implications for more than one subject area.

Conclusion

Modern lifestyle involves much more than only dietary alterations, e.g., physical activities, smoking and drinking habits and medication. As concerns diets, modern lifestyle does not necessarily imply a reduced energy intake but most likely increased use of refined food and an enlarged variety of raw materials. A lower energy intake will necessarily lead to a lower intake of vitamins if conventional food habits are adhered to. This will eventually result in impaired health in a significant number of individuals in various population groups. To avoid these consequences, food items with a higher nutrient density of vitamins should be identified to guarantee that physiological demands will still be covered.

Appendix 2

Report of the Minerals Working Group

Chairman: L. Hallberg
Rapporteur: A. Prentice
Members: B. M. Sandström, J. P. Mareschi, J. Durlach, K. Schmidt
Observers: E.R. Müller, M. Fondu

The impact of modern lifestyles in Europe on the supply of essential minerals was the focus for our discussion with special reference to Fe, Zn, Ca, Mg and Se. We accepted the basic premise that recent changes have taken place not only in the amount of energy consumed by individuals as a result of decreased energy expenditure accompanying a more sedentary way of life and a desire for a slimmer body image but also in the pattern and constitution of meals. In particular we noted the increasing use of commercially available ready-prepared meals both within the home and outside it in restaurants, schools, canteens, hospitals and other institutions.

Likely Effects of These Changes in Dietary Practice on the Supply of Specific Minerals

In many cases decreases in energy intake will be directly associated with decreased mineral consumption if the mineral density of the diet (mg/MJ) remains unaltered. For some minerals, for example Ca and Se, energy intakes *per se* are less relevant but changes in the consumption of specific food items are more important. Examples include calcium intakes which are highly dependent on the consumption of milk and milk products, and the beneficial effect of lean red meat in improving iron supply. In addition, the specific composition of meals is important as this can influence, in both positive and negative ways, the bioavailability of certain minerals. Examples here include the decreased availability of Zn and Fe when meals are rich in phytate and the increased absorption of iron when ascorbic acid is a constituent of the meal. Nutrient – nutrient interactions are likely to become more critical when the total amount of food consumed is reduced and when the proportions of these nutrients are markedly altered.

The committee expressed concern over the increasing use of self-medicated mineral supplements as we considered that these could increase the problem of adverse interactions between constituents of the diet. For example, evidence was given that calcium supplements could decrease the bioavailability of Fe, Mg and Zn.

Likely Outcome of These Changes in Mineral Supply

Functional disorders related to severe deficiency are well recognised for most minerals, for example Keshan's disease in areas of low Se intake, osteopenia in calcium deficiency and acrodermatitis enteropathica for zinc. However, at the present time we are unable to define, for any mineral, threshold intakes below which disease is likely. Indeed, critical levels of intake are likely to be very difficult to establish because of marked variations in individual requirements and the problem of inter-nutrient interactions between meal constituents. We also recognised the difficulties of determining actual intakes and absorption of minerals and other nutrients because of the limitations of currently available methods and data on the composition of foods. In particular we expressed concern that in many cases the composition of commercially-prepared ready-made meals was not available.

Lower mineral intakes are likely to increase the number of outwardly healthy individuals with marginal mineral deficiency. These marginal deficiency states might go largely unrecognised but problems would be expected to arise when the individual was stressed in some way, for example due to pregnancy or infection. Evidence was cited for the influence of marginal Fe, Mg and Zn status on pregnancy outcome, the roles of Zn, Fe, Se and Mg in immune defence and the ability of Se to detoxify the body after exposure to heavy metals. We considered that particularly vulnerable groups within our society were those involved in growth: the baby in the womb, infants, adolescents, pregnant and lactating mothers. An example here is the effect of marginal Fe status on mental function in children.

Although in principle we recognised that modern lifestyles are more likely to lead to increases in marginal mineral deficiency we also recognised the difficulty of defining and establishing marginal intakes. Biochemical indicators of status of varying degrees of reliability are currently used to assess status. The use of serum ferritin as a marker of Fe status is well established, for Se and Mg blood parameters such as lymphocytic Mg and blood concentrations of glutathione peroxidase and Se give useful information and there is hope that the new red blood cell metallothionine assay will prove effective for zinc. However, for some minerals, such as Ca, no reliable indices of marginal intakes are available. In addition, it is well known that in the short-term the body makes adjustments in the absorption and excretion of many minerals in response to fluctuations in supply but the extent to which long-term adaptations to decreased levels of mineral supply can occur is not known, especially in individuals already habituated to high mineral intakes. The committee considered that the likelihood of adaptation to low Fe intakes is small but that the potential exists for adaptation to occur for Ca, Zn and Mg.

Research Priorities

The committee were of the opinion that research priorities lay in three main areas:

1. Long-term, longitudinal studies within individuals are needed to understand the extent and nature of marginal deficiences, to determine the likely outcomes of changes in mineral supply and to determine the potential for adaptation. In particular more information is needed on the vulnerable groups such as children, adolescents and women of reproductive age.
2. More research is needed to define the relationships between biochemical and functional markers of status, mineral supply and outcome.
3. Improvements are needed in the reliability of mineral intake data. This will involve not only better assessments of how much people eat and the composition of foods but also determination of the effects of meal composition on nutrient–nutrient interactions.

Recommended Courses of Action to Lessen the Potential Problems associated with Recent Changes in Dietary Practice

The committee made the following recommendations:

1. People should be encouraged to increase their energy expenditure. This would be accompanied by increased energy intakes and enhanced mineral supply. In general it was considered that higher intakes of most minerals would minimize the possible problems of suboptimal mineral proportions and nutrient–nutrient interactions between meal constituents. In addition, increasing energy expenditure would have other health benefits, for example, to heart, muscle and bone.
2. Steps should be taken to improve the supply of minerals in the diet for those individuals unwilling to change their current lifestyle. There are a number of ways this could be done:

By increasing the mineral density (mg/MJ) of the diet. This could be achieved by encouraging the reduction in consumption of "empty calories" such as fat, sugar and alcohol. In certain circumstances diet fortification may be a practical solution. The committee believed that supplementation of an individual's diet was an option but one that should only be used for certain vulnerable groups, particularly pregnant women. For others, self-medication with mineral supplementation in all forms is not advisable and mineral intakes should be improved by sensible modification of dietary habits.

By altering meal composition to enhance mineral bioavailability and improve the quality of the diet. The committee recommended increased consumption of lean red meat, fruit, vegetables and less refined cereals. Food should be freshly prepared as far as possible to prevent deleterious effects of processing and storage on some minerals such as selenium and other nutrients involved in mineral supply such as ascorbic acid.

3. More nutrition education is required at all levels, particularly in the young. In addition, better information is required about food composition, especially of convenience foods, to enable both nutrition advisers and consumers to select foods which will optimize vitamin and mineral supply.

Subject Index

GPSR Compliance

The European Union's (EU) General Product Safety Regulation (GPSR) is a set of rules that requires consumer products to be safe and our obligations to ensure this.

If you have any concerns about our products, you can contact us on ProductSafety@springernature.com

In case Publisher is established outside the EU, the EU authorized representative is:

Springer Nature Customer Service Center GmbH
Europaplatz 3
69115 Heidelberg, Germany

Batch number: 10620881

Printed by Printforce, the Netherlands